The United States in the World-Economy

The United States in the World-Economy is a major new textbook survey of the rise of the United States within the world-economy, and the causes of its relative decline. With the USA being the dominant state in the contemporary world-economy, it is vital to understand how it got where it is today, and how it is responding to the current global economic crisis. Professor Agnew emphasizes the divergent experiences of different regions within the USA, and in so doing provides a significant 'new' regional geography, tracing the historical evolution of the USA within the world-economy, and assessing the contemporary impact of the world-economy upon and within it. No existing treatment covers the subject with equivalent breadth and theoretical acuity, and the guiding politico-economic framework provides a coherent radical perspective within which the author undertakes specific regional and historical analysis. *The United States in the World-Economy* will prove required reading for numerous courses in regional geography, area studies and the geography of the United States.

JOHN AGNEW is Associate Professor of Geography and Director of the Social Science Program at the Maxwell Graduate School, Syracuse University, New York. His previous publications include (as co-author) *Order and Skepticism* (1981) and *The City in Cultural Context* (1984).

Geography of the World-Economy

Series Editors:

PETER TAYLOR   *University of Newcastle upon Tyne* (General Editor)
JOHN AGNEW   *Syracuse University*
CHRIS DIXON   *City of London Polytechnic*
DEREK GREGORY   *University of Cambridge*
ROGER LEE   *Queen Mary College, London*

A geography without knowledge of place is hardly a geography at all. And yet traditional regional geography, underpinned by discredited theories of environmental determinism, is in decline. This new series *Geography of the World Economy* will reintegrate regional geography with modern theory and practice – by treating regions as dynamic components of an unfolding world-economy.

   Geography of the World-Economy will be a textbook series. Individual titles will approach regions from a radical political-economic perspective. Regions have been created by individuals working through institutions as different parts of the world have been incorporated in the world-economy. The new geographies in this series will examine the ever-changing dialectic between local interests and conflict and the wider mechanisms, economic and social, which shape the world system. They will attempt to capture a world of interlocking places, a mosaic of regions continually being made and remade.

   The readership for this important new series will be wide. The radical new geographies it provides will prove essential reading for second-year or junior/senior students on courses in regional geography, and area and development studies. They will provide valuable case-studies to complement theory teaching.

The United States in the World-Economy

A Regional Geography

John Agnew
Department of Geography, Syracuse University

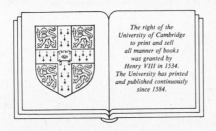

The right of the
University of Cambridge
to print and sell
all manner of books
was granted by
Henry VIII in 1534.
The University has printed
and published continuously
since 1584.

CAMBRIDGE UNIVERSITY PRESS
Cambridge
London   New York   New Rochelle
Melbourne   Sydney

Published by the Press Syndicate of the University of Cambridge
The Pitt Building, Trumpington Street, Cambridge CB2 1RP
32 East 57th Street, New York, NY 10022, USA
10 Stamford Road, Oakleigh, Melbourne 3166, Australia

First published 1987

Printed in Great Britain at the University Press, Cambridge

British Library cataloguing in publication data
Agnew, John
The United States in the world-economy: a
regional geography. – (A Geography of the
world-economy; v. 1)
1. United States – Foreign economic
relations
I. Title II. Series
337.73   HF1445

Library of Congress cataloguing in publication data
Agnew, John A.
The United States in the world-economy.
(Geography of the world-economy series)
Bibliography.
1. United States – Economic conditions.
2. United States – Economic conditions – Regional
disparities. 3. United States – Foreign economic
relations. 4. Regional economics. I. Title.
II. Series.
HC103.A37 1987   337.73   86–32743

ISBN 0 521 30410 5 hard covers
ISBN 0 521 31684 7 paperback

To Susan, Katie and Christine

Contents

Figures

Tables

Preface

The *New York Times* (August 27, 1985) says that 'Shoes are a "sunset" industry for advanced economies.' Similar comments have been made in the past few years about steel, cars and a variety of other major manufactured goods. This argument, to my knowledge, has never been applied to American agriculture. But the implication is that a law of economics indicates that all shoes worn by Americans should be made elsewhere, presumably in 'less-advanced economies.' 'High-tech' and service jobs are presumably the wave of the future in the United States.

Behind the prognoses of the experts, however, lies the pain of real people experiencing a major 'transition' as the United States shifts out of the comfortable era of growth and prosperity of the 1950s and 1960s. Many commentators believe that the country has entered into a period of decline. Its hegemony or dominance within the world is increasingly called into question. Some see this as a result of social rigidities and an overdemanding labor force draining the country's power and wealth (e.g. Olson 1982). Others are more inclined to search for the causes of relative decline in the context of America's changing relationship to the rest of the world. This is very much the position adopted in this book. America's present problems are viewed as the outcome of a process of *historical* involvement between the United States and an evolving world-economy (*économie-monde*) or global division of labor.

The history of modern social science is often seen as a struggle between 'universalizers' – those who presume that there exist universal laws applicable to all humans everywhere – and 'particularizers' – those who argue that no generalizations at all are feasible since everything is unique. This was very much to the forefront of arguments amongst geographers in the 1960s. The first group were self-defined 'scientific' geographers, who believed that the geographer's task is to discover the laws of spatial behavior. In practice this meant introducing space or distance as a variable into models borrowed from economics and psychology, the premier 'law-seeking' social sciences. The second group were self-defined 'idiographic' geographers, whose primary activity was mapping and otherwise describing places (countries, regions) in great detail.

What both schools missed was the possibility of a *via media*, a recognition that a significant level of abstraction is essential in explaining places, but that places can be different and distinctive. In a letter written in 1877 Marx captured the essential point (cited in Carr 1961: 82):

Events strikingly similar but occurring in a different historical milieu lead to completely different results . . . By studying each of these evolutions separately and then comparing them, it is easier to find the key to the understanding of this phenomenon; but it is never possible to arrive at this understanding by using the passe-partout of some universal historical–philosophical theory whose great virtue is to stand above history.

This is a powerful argument for a reinvigorated regional geography. Rather than generalizing from universal propositions about social process to a universal social form or social response, regional geography can provide a frame of reference for examining the relationship between causes and outcomes without the presumption of universality in outcomes.

One intellectual feature many universalizers and particularizers share is their acceptance of the boundaries between states as defining the fundamental unit of social science: the national 'society.' But, as Wallerstein (1984: 28) notes, 'this lumping together presumes what is to be demonstrated – that the political dimension is the one that unifies and delineates social action.' Rather, 'States are . . . created institutions' produced by and operating within a modern world-economy. This world-economy has, since coming into existence in the sixteenth century, acquired boundaries far larger than those of any single political unit. Within these limits, states and their capitalist producers compete economically and politically. The geographical localization of productive activities results from this.

The task of this book is to bring together the epistemological *via media* of regional geography with the historiographical challenge of a world-economy perspective through an examination of the involvement of the United States in the world-economy since its initial settlement by Europeans. It is an attempt both to define a 'new' regional geography and to develop a world-economy perspective sensitive to what Marx called 'historical milieu' or what geographers call place (Evans 1979; Smith, C.A. 1984).

The book is largely a synthesis of work done by others. A number of authors should be identified explicitly as major influences upon the perspective and argument: Immanuel Wallerstein, Alan Wolfe, William Appleman Williams, David Calleo, Richard Franklin Bensel, Nigel Harris, Peter Taylor, Wassily Leontief, Barry Bluestone, Bennett Harrison, Lynn E. Browne and Ricardo Parboni. Specific citations reveal specific influences. No one influence is predominant.

A number of people have been helpful in various ways in making this book possible. Peter Taylor initiated and encouraged the entire enterprise. Fred Shelley and Clark Archer provided useful critical comments and suggestions. D. Michael Kirchoff and Marcia Harrington drew the maps and diagrams. My colleague John Rees provided a number of useful references. Harriet Hanlon translated my handwritten draft into excellent typed copy. Finally, my wife Susan and daughters Katie and Christine tolerated my long hours working on the manuscript during the summer of 1985. A dedication of this work to them is a small reward for their persistent good humor and the relief they share with me that the book is finally finished.

John Agnew
Syracuse, New York

1

The United States in the world-economy

America has discovered the rest of the world. Over the past twenty years many Americans have become aware for the first time that the United States is not a world unto itself, economically self-sufficient and politically self-determining. To varying degrees they have discovered that their country is part of a world-economy, that their everyday lives are now vitally affected by the decisions and behavior of people elsewhere. Some have also developed a clear sense that the world-economy was largely dominated from 1945 until recently by Americans and the American economy. To them the discovery of the world thus reflects the reality of a threatened dominance. Americans can no longer take their insulation *or* their superiority for granted.

Between 1967 and 1973 two trends became clear. One was that the pattern of rapid US economic growth as it had been experienced since World War II had come to an end. Second, it became obvious that vaguely 'international' events, for example the military failure in Vietnam, rapid increases in the prices of basic raw materials (especially oil), military spending to counter the politico-military 'threat' from the Soviet Union, and the rise of foreign competition to American manufacturers (especially from Japan) within the domestic United States economy, could be implicated in the end of the golden age. Yet, by and large, acknowledgment of global interdependence has been limited by the persistence of modes of thinking that see international 'events' as separable and isolated from the workings of the national economy. They are 'shocks' rather than the product of routine interactions. Thus we have the Vietnam 'experience,' the 1974 oil price 'shock,' the Soviet 'threat' and the Japanese 'invasion.' Moreover, each remains the province of a different group of experts and commentators. In the tradition of the Indian fable of the blind men

1

and the elephant, each 'shock' is examined separately and left unrelated to the others.

What appear to be external shocks to the United States are in fact part and parcel of routine interactions between the US economy and the world-economy. This book examines these by tracing the historical evolution of the United States in the world-economy over the past three hundred years and by analyzing the contemporary impact of the world-economy upon and *within* the United States. This enables us to understand the global context of many current political and economic problems in the United States.

The world-economy

Many scholars now accept the need to examine patterns of economic and political development within countries in terms of the operations of a world-economy. These scholars approach the task from a variety of perspectives. Some, particularly those who focus solely on multinational corporations and the world monetary system in the 1970s and 1980s (e.g. Barnet and Müller 1974; Beenstock 1984; Calleo 1982; and Edwards 1985), tend to highlight the uniqueness of the contemporary situation. Others, and this study is an example, consider present-day global problems and issues in terms of global processes that have a long history (e.g. Wallerstein 1979; and Taylor 1985). This is not to say that there has been no qualitative change in the nature of these processes from the distant past to the present. Just as the world-economy itself evolved out of more localized economic systems and empires, so the world-economy has taken on different forms over time as it has become more and more integrated, covering ever wider geographical areas and more and more activities (resource extraction, capital investment, trade in manufactures, services, etc.). The workings of the modern world-economy, then, can only be understood in a historical–geographical framework.

Wallerstein is the major contemporary proponent of this viewpoint and his work is both the background to, and the inspiration for, the present study. Wallerstein (1979: 17) argues that at one time all societies were 'minisystems': 'A minisystem is an entity that has within it a complete division of labor, and a single cultural framework.' Such systems are found only in very simple agricultural or hunting and gathering societies. Such minisystems no longer exist in the world: 'any such system that became tied to an empire by the payment of tribute as "protection costs" ceased by that fact to be a "system".'

Then there came world-systems, 'units with a single division of labor and multiple cultural systems. It follows logically that there can be two varieties of such world-systems, one with a common political system and one without.' The former are called 'world-empires,' and the latter 'world-economies.' Until the advent of capitalism in Europe, world-economies were unstable and tended towards 'disintegration or conquest by one group and hence transformation into a world-empire. Examples of such world-empires emerging from world-economies are all the so-called great civilizations of premodern times, such as China, Egypt, Rome.'

World-empires undermined the economic dynamism of their territories by using too much of their surpluses to maintain their bureaucracies. Around AD 1500 a new type of world-economy, the modern capitalist one, began to take form:

In a capitalist world-economy, political energy is used to secure monopoly rights (or as near to them as can be achieved). The state becomes less the central economic enterprise than the means of assuring certain terms of trade in other economic transactions. In this way, the operation of the market (not the *free* operation but nonetheless its operation) creates incentives to increased productivity and all the consequent accompaniment of modern economic development. (Wallerstein 1974: 16)

The causes of capitalism's success as a means of organizing a world-economy are complex, but two are fundamental. New transportation technology allowed far-flung resource areas to be connected with markets, and European military technology provided the power to enforce favorable terms of trade. Released from the burden of maintaining relations of tribute within their economic zones, merchant capitalists could attend to the expansion of their interests. Of special importance were English and Dutch capitalists, who were able in the sixteenth century to beat back the Spanish–Habsburg attempt to turn the emerging world-economy into a world-empire. After that, capitalism spread throughout the globe even though it was periodically threatened by conversion into a world-empire dominated by one or other hegemonic state (Wallerstein 1974). By 1900 it was truly global.

Wallerstein identifies four basic elements in the capitalist world-economy. First, the world-economy consists of a single world market. Within this, production is for exchange rather than use: producers exchange what they produce for the best price they can get. As the price of a product or commodity is not fixed but set by the market there is competition between producers. More efficient producers can undercut other producers to increase their share of total production and achieve monopoly control. In recent years the world market has been dominated by multinational corporations.

There is controversy over whether the world-economy can properly be labelled 'capitalist' before the late eighteenth century. Only since that time has the world-economy provided a price-fixing market. Previously, the argument runs, the world-wide movement of commodities took place without labor becoming a commodity. Thus, there may have been a world-economy but it was not as yet a capitalist one. This need not detain us here since most commentators seem agreed that between 1500 and the late eighteenth century there was a vast expansion of trading relations between Europe and other parts of the world. The major point is that there was a *qualitative* change in the nature of the world-economy in the late eighteenth century associated with European industrialization. It is from this more recent time that the self-regulating world market dates (see Brenner 1977).

Second, in the modern world-economy there has always been a territorial division between political states. This division both pre-dates and grew along with the spread of the world-economy. It ensures that the world-economy cannot be transformed into a world-empire. At the same time it provides for the protection of developing industries and a means for groups of capitalists to protect their interests and distort the market if it is to their advantage. The major result of this process is a competitive state system in which each state attempts to the best of its ability to insulate itself from the rigors of the world market while attempting to turn the world market to its advantage.

Third, the modern world-economy has established a basic 'three-tiered' geography as it has expanded to cover the globe. This geography is defined by the international division of labor at a particular time. The initial world-economy consisted of Europe and those parts of South and Central America under Spanish–Portuguese control. The rest of the world was an 'external arena.' Spanish activity in America was fundamental in forming the world-economy. Later, by means of plunder, European settlement in place of aboriginal groups and the reorientation of local economies to the world-economy, the rest of the world was transformed from an external arena into a periphery. The world-economy thus came to consist of a *core* (Western Europe at first, later joined by the United States and Japan) and a *periphery*. The core is defined in terms of processes in the world-economy that have led to relatively high incomes, advanced technology and diversified production, whereas a periphery is defined in terms of processes leading to low incomes, primitive technology and undiversified production. The core *needs* the periphery to provide the surplus to fuel its growth. Uneven development, therefore, is not a recent phenomenon or a by-product of the world-economy; it is one of the modern world-economy's basic components.

There is movement between the two categories of core and periphery as attested to by the 'rise' of the United States and Japan. A third geographical tier, the *semi-periphery*, applies to the processes operating in certain parts of the world to provide movement between periphery and core (and vice versa). These are zones in which a mix of core and peripheral processes are at work. Essentially, political conditions determine whether core processes come to dominate and allow 'upward mobility' within the world-economy. In particular, the timely application of protectionism and other autarkic measures can allow for development but only in conjunction with favorable global economic conditions. Wallerstein emphasizes the importance of the third tier, the semi-periphery. Not only do economies in this group stand between the core and the periphery, some may eventually fall into the periphery, as Spain did in the seventeenth century, and others may eventually rise into the core as the United States did in the nineteenth century. The semi-periphery thus provides a geographical dynamism to the world-economy (Agnew 1982).

Fourth, and finally, the modern world-economy has followed a temporally cyclical pattern of growth and recession. The causes of this pattern are the subject of controversy (Kindleberger 1978). But there is now considerable evidence that, at least since the eighteenth century, the world-economy has gone through four major cycles of growth (A) and stagnation (B) (Kondratieff 1984). These have been identified in time–series data for a wide range of economic phenomena (approximate dates):

I: $1780-90 \rightarrow A \rightarrow 1810-17 \rightarrow B \rightarrow 1844-51$
II: $1844-51 \rightarrow A \rightarrow 1870-5 \rightarrow B \rightarrow 1890-6$
III: $1890-6 \rightarrow A \rightarrow 1914-20 \rightarrow B \rightarrow 1940-5$
IV: $1940-5 \rightarrow A \rightarrow 1967-73 \rightarrow B \rightarrow ?$

A widely accepted, if not definitive, explanation of these regular fifty-year cycles (often called Kondratieff cycles after the Russian economist who first identified them) focuses on the contradiction between the short-run interests of individual firms, on the one hand, and the long-run collective needs of business as a whole, on the other. In times of growth profits are higher and firms tend to overinvest in production and new technology. With no central control over production this leads to overinvestment and overproduction. Stagnation follows.

Table 1.1. *A model of dominance and inter-state rivalry*

	Britain	USA
	1780–90	1890–96
A₁ Ascending hegemony	Rivalry with France. Industrial revolution.	Rivalry with Germany and Britain. Mass production.
	1810–17	1914–20
B₁ Hegemonic victory	Commercial dominance in Latin America and India. 'Workshop of the World'.	Commercial dominance when free trade system collapses. Decisive defeat of Germany.
	1844–51	1940–5
A₂ Hegemonic maturity	Era of free trade: London (The City) is the financial center of the world-economy.	Bretton Woods financial system based on US dollar: New York (Wall Street) is the financial center of the world.
	1870–5	1967–73
B₂ Declining hegemony	Classic age of imperialism as United States and European powers challenge Britain. 'New' industrial revolution outside of Britain.	Reversal to protectionism. Use of US financial hegemony to maintain prosperity. 'New' industrial revolution in Japan and East Asia.

The correlation with technological change is high. Growth or A-phases are associated with major periods of technological innovation. For example, the first A-phase is the period of the original 'industrial revolution' in England. Subsequent periods fit the pattern as well, with railways and steel in II A, chemicals and electricity in III A, and aerospace and electronics in IV A.

But there are also important political correlates. Wallerstein *et al.* (1979) postulate that competition between states involves the geographical *redistribution* of cyclical effects to the benefit of some strong states and to the cost of other weaker ones. Thus, states can differentially exploit or suffer from cyclical shifts depending on their productive efficiency, commercial supremacy and ability to restrict competition from rivals (Bousquet 1980). According to Wallerstein *et al.* (1979) the four Kondratieff cycles, when put in a political context, can be described as two 'paired Kondratieffs' (see Table 1.1). The first pair covering the nineteenth century involves the rise and demise of British dominance and the second pair describes a similar trend for the United States in this century.

No inevitability should be read into this pattern of paired Kondratieffs. In particular, this should *not* be seen as a case of 'history repeating itself!' Rather, there is a *similar* pattern to the ways in which first Britain and then the United States gained and then lost pre-eminence. The extent and nature of the dominance exerted by each has been different. For example, Britain followed a 'mixed' strategy of formal and informal imperialism whereas the United States has largely eschewed the territorial control outside its own borders required by formal imperialism. Moreover, US dominance has generally involved investment overseas and manipulation of the global monetary system rather than the raw-materials and markets bias of the British.

More importantly, the US hegemony has been constantly challenged by the existence of, and military competition from, the Soviet Union, a state representing a different image of world order. British dominance in the nineteenth century never faced such a challenge.

Finally, American hegemony has been achieved through an internationalization of the world-economy to an extent unknown in the eighteenth and nineteenth centuries. Large American-based multinational corporations have been major instruments of American hegemony. But *they*, as much as other states such as the Soviet Union or Japan, have now become the major threat to American hegemony. This seeming paradox is explored in a later section of this chapter and provides a major theme for the book as a whole.

A brief addendum on cycles: before 1780 there is little evidence for Kondratieff cycles. So it is dangerous to project backwards the dynamics of the world-economy from more recent times. However, there is some support for what are called 'logistic' waves of three hundred years or so. The two logistics Wallerstein has focused on are as follows:

$$c. 1050 \rightarrow A \rightarrow c. 1250 \rightarrow B \rightarrow c. 1450$$
$$c. 1450 \rightarrow A \rightarrow c. 1600 \rightarrow B \rightarrow c. 1750$$

The first logistic covers the rise and decline of feudal Europe. The second covers the emergence and then the crisis of the European world-economy based upon agricultural capitalism. It is in this second logistic that the modern world-economy first emerged and the groundwork was laid for later periods of growth and stagnation and the dynamics of inter-state rivalry and hegemony indicated by the four Kondratieff cycles of the past two hundred years.

To summarize the argument on the world-economy, since the sixteenth century the logic of the world-economy has spread progressively over the globe. Today no single country can escape its impact. As should be already apparent, the United States is no exception. Its history and that of the world-economy are interwoven. This book is an exploration of that relationship. But we first must examine why there might be resistance to thinking this way, particularly about the United States.

We are the world? The problem of national exceptionalism

The United States was created by a massive invasion of North America by Europeans. This movement displaced or destroyed all the existing societies or 'minisystems' in its path. It is emblematic that these local societies should have succumbed to a population actively engaged in the making of the world-economy. Of course, this process was rarely understood as such at the time or since. More typically, the penetration of North America from the Atlantic coast was seen as a crusade or mission. Robertson describes this as follows:

The American sense of uniqueness has come from the belief that the mission of its people was to create a nation where a nation did not exist. Nationalism included expansion, but it was expansion into the wilderness which was *part* of the nation and at the same time *had to become* part of the nation. So Americans were crusaders, bringing civilization and freedom to the wilderness. The crusade was unique; it took place in a New World, and it created one. (Robertson, J. O. 1980: 26)

The American case for exceptionalism is often most persuasive. Its major aspect is its focus on the unique *origins* of the American state. Robertson (1980: 26) characterizes this as follows in his discussion of the major myths in American political culture:

Americans are a new people, formed out of a migration of people seeking freedom in a new world. The nation was founded in a revolution which was both the first war of liberation and the first lasting overthrow of an *ancien régime*. That revolution created a *new* nation dedicated to the spread of freedom and democracy and equality. The history of that people and nation has been the struggle, physically and geographically as well as morally and ideally, to spread freedom across the continent and around the world.

Rather than having a 'natural,' stable, established territorial base, then, the United States as a complete nation had to be captured by territorial expansion. This 'destiny' or 'mission' has had several justifications. In the early years of the Republic geographical determinism was invoked. Areas contiguous to the established area of settlement were thought of as natural extensions to national territory. In line with this, 'the young expansionists who led the country into war in 1812 in the hope of conquering Canada and Florida appealed to the God of Nature on behalf of their plans' (Pratt 1935: 36). One of them asserted: 'In point of territorial limit, the map will prove its importance. The waters of the St Lawrence and the Mississippi interlock in a number of places, and the great Disposer of Human Events intended those two rivers should belong to the same people' (quoted in Pratt 1935: 336–7). Others issued similar declarations on behalf of the Author of Nature or some such entity.

The dominant justification, however, became the providential mission of 'America' to spread American ideals and institutions to the Pacific – and beyond. This mission began to prosper during the Jacksonian era when zealots began to write of American 'Manifest Destiny' and 'The Great Nation of Futurity.' In 1847 the Secretary of Treasury placed in his report a section referring to the aid of a 'higher than any earthly power' which had guided American expansion in the past and which 'still guards and directs our destiny, impels us onward, and has selected our great and happy country as a model and ultimate centre of attraction for all the nations of the world' (quoted in Pratt 1935: 343).

A third, and today rarely articulated, argument for American destiny, was racial. Both before and after the publication of Darwin's *Origin of Species* there were those in Congress and elsewhere who argued that the 'Anglo-Saxon' population of the United States was destined to flourish while other populations either withered away or were progressively amalgamated with the superior 'race.' Before Darwin added an air of authority to such ideas the emphasis was placed upon the superior fecundity of American Anglo-Saxons; after Darwin the concept of superior fitness was invoked. John Fiske, a leading advocate of the fitness argument, believed that Anglo-Saxons (and other 'Teutons') in the United States had evolved beyond other races by dint of their superior political principles, growth in numbers and economic power. Fiske was ambitious. 'The day is at hand,' he wrote, 'when four-fifths of the human race will trace its pedigree to English forefathers, as four-fifths of the white people of the United States trace their pedigree today [1885]' (quoted in Pratt 1935: 348). As Pratt (1935: 34) notes: 'This was surely encouraging doctrine to Americans or British who wanted an excuse to go a-conquering.'

A final justification, though apparently venal, was to be seen in a providential light: the economic value of new territory. Thus the Spanish–American war was widely advertised as a civilizing as much as a commercial affair. To the *American Banker* of New York in 1898 the coincidence between victory over Spain in the Philippines and the attempts of European powers to subdue China and monopolize its markets had 'a providential air' (quoted in Pratt 1935: 353). President McKinley's appeal for divine guidance as to the future of the Philippines is a fitting summary of this religious–economic justification for national exceptionalism: 'One night it came to me this way – I don't know how it was but it came ... that we could not turn them over to France or Germany – our commercial rivals in the Orient – that would be bad business and discreditable' (quoted in Pratt 1935: 353).

Up to and beyond continental boundaries, then, the political definition of the United States has been viewed in exceptionalist terms as the outcome of a unique, often provident, place 'under the Sun.' At every step in the process of American territorial and economic expansion there have been Americans ready to give the best of reasons, both pious and fatalistic, for its necessity. More recently, American exceptionalism has been restated most definitively in terms of America as the 'homeland of liberty' (Merk 1963). Merk sees manifest destiny and expansionism as temporary aberrations in a flow of history otherwise dominated by the Idea of Mission: spreading 'democracy' or saving it from its foes. Huntington, in criticizing the 'moralism' of American foreign policy during the Carter administration for its neglect of 'enforcement' – liberty, it seems is best spread by 'troop deployments, carrier task forces, foreign aid missions and intelligence operatives' (Huntington 1982: 34) – writes that 'The United States has no meaning, no identity, no political culture or even history apart from its ideals of liberty and democracy and the continuing efforts of Americans to realize those ideals' (Huntington 1982: 36). He quotes approvingly from a Yugoslav dissident in defense of American exceptionalism:

The United States is not a state like France, China, England, etc., and it would be a great tragedy if someday the United States became such a state. What is the difference? First of all, the United States is not a national state but a multi-national state. Second, the United States was founded by people who valued individual freedom more highly than their own country.

And so the United States is primarily a state of freedom. And this is what is most important. Whole peoples from other countries can say, Our homeland is Germany, Russia, or whatever; only Americans can say, my homeland is freedom. (Quoted in Huntington 1982: 3–35)

If the focal point of American exceptionalism is seen to lie with the Idea of America as the 'homeland of freedom,' its realization is based upon a Tocquevillian scenario of 'free' farmers struggling to settle

'free' land in fulfillment again of an Idea: the West as a yeoman's garden. It is the frontier to the west that epitomizes for most exceptionalists the social context for the 'American dream' of a prosperous
world safe from the tragedies of life in nineteenth-century Europe:
'Fertile soil on the high plains, open spaces, seemingly "virgin" lands
beckoned the independent yeoman Jefferson had celebrated as
America's best hope, and seemed an assurance of permanent tranquility' (Trachtenberg 1982: 20). It is the unique American experience
of the 'seed of liberty' planted in the fertile garden of the western
frontier, then, that is held to account for American exceptionalism.

There have been numerous attempts to displace the rhetoric of
exceptionalism from its historiographic pedestal. Some of the so-
called 'progressive' historians, for example, questioned the idealist
and supernatural arguments upon which explanation of the American
Revolution was preponderantly based in nineteenth-century historical
writing. Instead, they sought to explain the Revolution and the
development of the United States in terms of socio-economic relationships and interests. Rather than a providentially inspired filling-out of
the national space, the continental expansion of the United States is
seen by historians in this tradition as either empire-building or
annexation in pursuit of wealth by politically dominant elites. More
recently, some so-called 'revisionists' have added the *ideology* of
exceptionalism as an important cause in its own right of expansion
(e.g. Williams 1969; Drinnon 1980; Rosenberg, E. S. 1982; and
Trachtenberg 1982). The attraction of these viewpoints is overwhelming if one finds the transcendental idealism of exceptionalism
unappealing as an approach to historical explanation.

The American Revolution is a good starting-point. In particular, is
it exceptional? After a studied examination of competing claims one
investigator reaches the following conclusion: 'If we were to confine
ourselves to examining the Revolutionary rhetoric alone, apart from
what happened politically or socially, it would be virtually impossible
to distinguish the American Revolution from any other revolution in
modern Western history' (Wood 1966: 25–6). But as he adds later:

The rhetoric of the Americans was never obscuring but remarkably revealing
of their deepest interests and passions . . . The ideas had relevance; the sense
of oppression and injury, although often displaced onto the imperial system,
was nonetheless real. It was indeed the meaningfulness of the connection
between what the Americans said and what they felt that gave the ideas their
propulsive force and their overwhelming persuasiveness. (Wood 1966: 31)

It was the ideas and action of people forged under the circumstances of a revolutionary situation, therefore, that created the Revolution, rather than the Idea of Liberty working superorganically. It is not, then, that ideas are without consequences but that they do not descend from an Idea and they do develop in socio-economic contexts. Similar contexts bring forth similar ideas:

In the kinds of ideas expressed, the American Revolution is remarkably similar to the seventeenth-century Puritan Revolution and to the eighteenth-century French Revolution: the same general disgust with a chaotic and corrupt world, the same anxious and angry bombast, the same excited fears of conspiracies by depraved men, the same utopian hopes for the construction of a new and virtuous order. (Wood 1966: 26)

After the Revolution, however, the idealist vision of America as a 'land of exceptional virtue' is still more vulnerable. Even in its early years the young American Republic was vigorously expansive – at others' expense. Even Jefferson, often portrayed as the most virtuous of the early American political leaders, was not without a vision of an American empire (Pratt 1935: 336; Lafeber 1963: 3). The War of 1812 resulted from American territorial ambitions in Canada and Florida (Pratt 1957). More generally, the 'ideals' of the American Revolution did not exist in a political and intellectual vacuum:

Americans might dislike traditional diplomacy and power politics, but they no longer viewed them as feeble structures which would fall at the first blowing of the trumpets of liberty. Americans had become aware that, behind the forms of foreign policy and diplomacy as they existed in Europe, there lay a coherent system of thoughts, of principles, of methods. They had begun to use its terms and to think in its concepts. (Gilbert 1961: 89)

But it is the mechanism upon which the perpetuation of American exceptionalism is based, the 'Arcadian frontier,' that is most easily disposed of. Besides wresting the 'free' land from its previous possessors (native Americans and Mexicans), the continental expansion of the United States procured a resource base unsurpassed by any of the nineteenth century's other imperialisms. A national policy of conquest, settlement and exploitation was at first geared towards agricultural settlement, but by the Civil War, and perhaps as both cause and consequence of it, this policy had become much more of a 'massive industrial campaign' (Stedman Jones 1973; Trachtenberg 1982: 20). By the 1870s the United States was undergoing one of the most rapid transformations from an agrarian to an industrial society ever experienced. The 'West' rapidly became an integrated part of this new order:

Following the lead of the railroads, commercial and industrial businesses conceived of themselves as having the entire national space at their disposal: from raw materials for processing to goods for marketing. The process of making themselves national entailed a changed relation of corporations to agriculture, an assimilation of agricultural enterprise within productive and marketing structures . . . Agricultural products entered the commodities market and became part of an international system of buying, selling, and shipping. (Trachtenberg 1982: 20–1)

As for the society of family farmers and small homesteaders upon which the image of the 'Arcadian frontier' rests, this became increasingly fictive in large parts of the West as the nineteenth century progressed. Following the Homestead Act of 1862, which offered 160 acres of the public domain to individuals for the nominal fee of $10, the hope was otherwise. But between 1860 and 1900 only one-tenth of the new farms established in the United States were acquired under the Act (Shannon 1945). Moreover, huge 'land grants' to railroad companies and the incredible growth of wheat and cattle 'enterprises' owned by banks and meat-packing firms in the Northeast and Midwest created economic conditions in which the 'freedom' of family farmers was severely circumscribed:

By the 1890s, food production and processing had joined mining as a capital-intensive, highly mechanized industry. The translation of land into capital, of what once seemed 'free' into private wealth, followed the script of industrial progress, however much that script seemed at odds, in the eyes of hard-pressed farmers, with the earlier dream. (Trachtenberg 1982: 23)

By the 1890s the United States had, in the view of many influential contemporaries, fulfilled its 'continental destiny.' The time was ripe, they believed, for a commanding world role. The kernel of this perspective was stated by Alfred Thayer Mahan, Henry and Brooks Adams, Henry Cabot Lodge, John Hay and Theodore Roosevelt (Vidal 1986). This group not only pressed for overseas expansion but developed what might be called an imperialist world view akin to that popular in Britain at that time. Given the economic crisis of the early 1890s their arguments began to take on a sense of urgency that they had previously lacked (Calleo and Rowland 1973).

To Turner, the contemporary historian most committed to the idealized 'frontier' as the key to the American difference, the 'internal' frontier was gone and new frontiers had to be pursued. But he stressed that this was not a new departure in the national development of the United States. Other frontiers, for example in Latin America, had long been pursued. In one paragraph of his famous article 'The Problem of the West', Turner brought together his view of the relationship between the closing internal frontier and an invigorated American foreign policy:

For nearly three hundred years the dominant fact in American life has been expansion. With the settlement of the Pacific Coast and the occupation of the free lands, this movement has come to a check. That these energies of expansion will no longer operate would be a rash prediction; and the demands for a vigorous foreign policy, for an interoceanic canal, for a revival of our power upon the seas, and for the extension of American influence to outlying islands and adjoining countries, are indications that the movement will continue. (Turner 1896: 289)

American activity in the Far East 'to engage in the world-politics of the Pacific Ocean,' the 'extension of power' and 'entry into the sisterhood of world states', were not sudden and inexplicable. They were 'in some respects the logical outcome of the nation's march to the Pacific' (Turner, quoted in Lafeber 1963: 71).

As in the European countries, a vision of empire took shape in the United States during the nineteenth century. Though justified transcendentally as the pursuit of an 'American dream' unlike mere dreams, this vision grew from primitive roots. In a world of 'sovereign' states into which the United States was born the choice was clear: own or be owned, expand or be consumed. Ultimately, and this was clear from the beginning, prey was also predator.

But the image of national exceptionalism still persists as a major mental barrier to seeing the United States as part of the world-economy. The American 'experience' is often still seen as so different from that of other parts of the world to the extent that it is not of this world at all.

Obviously, many Americans have thought that they were engaged in something qualitatively different and superior to European 'empire-building.' But what is important to remember is that behind the enthusiasm that this inspired, important as it was, lay a set of connections to, and a competition of interests with, the places on the other side of the Atlantic. The settlement along the Atlantic coast of North America that became the territorial–economic core of the United States was a product of European commercial–political expansion in the seventeenth and eighteenth centuries. American independence from Britain was the result of the breakdown of routinized trans-Atlantic relations brought on by the British revenue legislation of 1764 (Dickerson 1951) and, after Independence, America remained materially and ideologically tied to Europe (Gilbert 1961). So, altogether, the history of the United States cannot be separated from that of the world-economy.

However, it is important to stress that this does not involve us merely in some deductions from a universal history. Every place and state has its own particular and peculiar relationship to the evolution of the world-economy. From this point of view, the ideology of American exceptionalism is an integral part of American history (Harrington 1986). Furthermore, America was settled by Europeans as it was incorporated into the world-economy. It also became the first 'settler-state' after achieving its political independence from Britain. Within its original territory along the Atlantic coast it also contained contrasting, and ultimately incompatible, modes of socio-economic organization: a plantation agriculture based on slavery in the South and classic capitalist or 'free' enterprise in the North. These and other particularities, for example the lack of foreign invasion and, until Vietnam, costly 'colonial' wars, have continued to interact with the 'shaping' of America by the world-economy since the late eighteenth century to create American history. American history, therefore, *is* different, but it is also critically dependent on America's interactions with the world-economy.

The American impasse

The reality of American interdependence with, and vulnerability to, the world-economy has perhaps only become clear to many Americans since 1967. Until that time the American economy seemed invincible and, at least since 1945, without peer. It is interesting that it has been since 1967 that most 'non-exceptionalist' writing about the American economy and polity has appeared. But the very features of the US economy that were most responsible for growth in the period 1945–67 – economic concentration, growing state intervention, and expansion overseas – had by the late 1960s become liabilities (Wolfe 1981a: 24). The end of what seemed to be a 'permanent boom' created what Wolfe has termed 'America's impasse' – what is good for America is no longer good for the world-economy, and vice versa. Until the late 1960s there seemed to be an identity between the United States and world economies. The world-economy, like history, was on our side. Wolfe (1981a: 143) elaborates:

What is good for the nation-state is not necessarily good for the world, and vice versa. Protecting the former, the United States incorporated the national interest into its international activity, but the hegemony over the world economy that made such an eclectic approach possible was undermined as the approach itself began to work. As that took place, the United States either had to suppress growth abroad in order to maintain its advantage, or to channel growth into American-based multinational companies. Both were tried, yet there was no way that the fantastic American advantage over the rest of the world could be maintained forever.

Three elements compose the underpinnings for America's boom – and bust – in the post-1945 world. They also provide the constituents of a *growth coalition* of big business leaders, labor leaders and politicians that emerged into political dominance in the United States after World War II (Wolfe 1981a). The first is economic concentration. In almost every American industry the control over the market came to be exercised by one or few firms. The proportion of total manufacturing assets held by the 200 largest corporations increased from 45 percent in 1947 to 60.4 percent in 1968 (Blair 1972: 331). In the 1960s a new style of concentration using takeovers of smaller firms began, reflected in the growth of conglomerates such as ITT, Gulf and Western, and United Technologies. These giant firms were active in expanding abroad.

Expanding concentration was accompanied and encouraged by the growth of American government. Government expenditures, which were 12.4 percent of the gross national product (GNP) in 1940, reached 24.6 percent in 1950 due to wartime expansion, increased to 27.8 percent in 1955, then to 28.1 percent in 1960 and to 30 percent in 1965 (Mandel 1975: 487). This spending went disproportionately to military production to meet the long-term Soviet 'threat,' which in turn was concentrated overwhelmingly in the hands of a few large firms. Only since the 1960s have non-military social expenditures constituted a large segment of American public expenditures.

These trends were also reinforced by the activities of US corporations overseas. Most of these activities, with the exception of the oil industry, were located in industrially advanced countries and involved direct investment by the largest corporations (Musgrave 1975: xi). In the short run this benefited the American economy in the form of repatriated profits. But in the longer run and by the late 1960s, as domestic technology and management followed capital investment, traditional exports were replaced by foreign production by US affiliates to the detriment of employment in the United States (Musgrave 1975).

The three components came together to achieve a peak of growth in the early 1960s. The growth was astonishing. In constant 1958 dollars the American GNP increased from $511.7 billion in 1961 to $636.6 billion in 1965 and $716.5 billion in 1968. Industrial production nearly doubled during the decade of the 1960s. New construction, especially in suburban housing, and new investment in new factories and equipment more than doubled (Wolfe 1981a: 33–4). Unemployment was at all-time lows, 5.7 percent in 1963 and 3.8 percent in 1966. Inflation was only 2 percent in 1965 and 3.8 percent in 1966.

But in 1967 and 1968 inflation and unemployment went up together. 'Stagflation' set in. The early 1970s were marked by huge increases in trade deficits (only exceeded by those of the early 1980s), which in combination with currency speculation would bring about in 1971 the end of the US-dominated Bretton Woods system of international monetary regulation. The post-war wave of economic growth was over (Wolfe 1981a: 38). What had gone wrong?

As long as the US economy was growing, economic concentration appeared to pay dividends – expanding profits and providing higher incomes that stimulated consumption. But big firms, protected from competition, failed to engage in sufficient research and development, which led to declines in productivity. Built-in wage increases for workers in many large firms also cut into profits. Corporate control over pricing ensured that prices did not fall as fast as general economic conditions. This increased inflation.

The growth of government became a drain rather than a stimulus. Huge sums were still required to pay for the Vietnam War, to maintain a vast US military presence around the world to 'contain' the Soviet Union, and by the early 1970s to pay for the social policies enacted in the 1960s to meet the demands of groups in the United States who were not sharing in the general prosperity of that time. All this had to be done without raising taxes, a political taboo given the hostility of different groups to one or other variety of government spending. The way out was to increase the money supply and encourage consumption through private indebtedness, thereby increasing inflation.

As first cause and then consequence of the domestic economic slowdown, American multinational corporations found overseas investment more profitable than investments in the United States. Between 1960 and 1970 new corporate investments abroad expanded from 21 percent to 40 percent of total investments (Castells 1979: 107). In the banking sector, overseas deposits increased from 30 percent of total deposits in 1965 to 70 percent in 1972 (Castells 1979: 107). Not only did this increase unemployment in the United States, but it also reduced the ability of American manufacturers to compete globally as investors turned abroad. Altogether, therefore, economic concentration, the growth of government, and overseas expansion, *reduced* the prominent position of the United States within the world-economy just as they had once worked to bring this prominence about. As

Wolfe (1981a: 41) puts it: 'An American public that had once received an imperial dividend was now being asked to pay an imperial price.'

Since the late 1960s successive American governments of both major political parties have struggled to find a way out of the impasse, to get back on the growth track. None has yet succeeded. One strategy has been to protect American industries from foreign competition through increasing tariffs and negotiating 'voluntary quotas.' But this runs the risk of 'competitive protectionism' and is opposed, for obvious reasons, by American-based multinationals. Another strategy has been to manipulate the global monetary system to America's advantage. The relatively 'closed' nature of the US economy (it has a very low import–GNP ratio of 0.15) enables the United States to undercut its competitors by devaluing *its* dollar, the metric of world commerce. From 1971, when the US government unilaterally ended the convertibility of the dollar into gold by abrogating the Bretton Woods Agreement of 1944, until 1978, US governments financed trade deficits with payment in US currency (Parboni 1981: 41). Between 1970 and 1978 the United States ran up a current account deficit of more than $30 billion, essentially by printing more dollars.

Since 1978, and especially from 1981 until 1985, massive US government deficits provided a third 'way out.' The US government increased expenditures without commensurate increases in revenues (Hershey 1986). But deficits *must* be financed at whatever cost; the US government, by raising interest rates, attracted huge volumes of foreign investment into the United States. This had the net effect of strengthening rather than devaluing the dollar. In the short run this reduced inflation in the United States. But a strong dollar weakens American manufacturing by making foreign goods relatively cheaper, inside and outside the United States. So the long-run picture is bleak, particularly for US manufacturing industry. The dilemma is as follows: although a lower dollar value would ease the country's trade deficit and help manufacturers, the 1985 US federal budget deficit of $200 billion *requires* a strong dollar to keep America attractive to foreign investors (Schneider 1985). The US budget deficit is a consequence of a massive increase in the rate of military spending and the increasing cost of social programs such as social security (pensions) without there being any accompanying increase in taxes. Indeed, the Reagan administration *cut* income and corporate taxes in 1981 on the assumption that the money left in private hands would lead to a national investment boom. That has not yet occurred. To the contrary, the tax cut did not generate enough domestic savings to service the US budget deficit let alone generate an investment boom (McIntyre

1986; Rosenbaum 1986). So the third strategy of attracting foreign investment to finance vast government deficits and thus stimulate the US economy and reduce inflation seems as unlikely a *long-run* solution to the American impasse as the other two (Modigliani 1985). And as Europeans are all too well aware, to work it depends on permanent recession-like conditions in other industrialized countries. America's investment gain has been their loss. How long can this continue?

In 1986 American joint efforts with other countries to lower the value of the dollar marked the beginning of a combination of the 1971–8 and 1978–85 monetary strategies. Interest rates dropped but the central problem of financing US trade and federal deficits remained. In 1985 the US current account deficit (which includes trade in merchandise and American investments overseas) surged to $117.7 billion and the federal deficit stood at $200 billion.

The impasse has, however, already had dramatic effects on the internal geography of the United States, too. Often labelled the emergence of a 'post-industrial' America (e.g. Clark, D. 1985), this involves the large-scale 'shake-out' of entire sectors of US manufacturing industry, such as steel, rubber, chemicals, textiles and a wide range of consumer goods, and the growth of service industries and some 'high-tech' manufacturing industries as replacements. The problem is that many of the new industries are locating in different regions of the country from the old ones, giving rise to what is called the 'decline' of the traditional manufacturing belt in the Northeast and the 'rise' of the so-called 'Sunbelt' in the South and West (Sawers and Tabb 1984). American businesses now search the United States, as they do the globe, in pursuit of more 'profitable' locations. These are ones in which costs from wage rates, taxes, pollution regulations and the like can be reduced.

For the American economy, therefore, free trade is no longer what it was claimed to be in the years of certain hegemony. It is an increasingly costly philosophy of trade (Kuttner 1983). In the long run it can only lead to a decline in the standard of living of most Americans – particularly those living in the northeastern United States. But protectionism is now equally problematic. So much of 'American' business operates on a global scale or beyond the borders of the United States that without concerted action between the United States and other states little can be done to restrict its operations. In the absence of such restrictions – and it would require the United States to abandon its claims to hegemony for it to happen – US governments have been left with one course of action: their ability to manipulate the global monetary system because of the centrality of the US dollar to world commerce. As one Italian economist is reported to have said in response to a question from an American about the condition of the Italian economy in 1985: 'Our situation is similar to that in the United States, except that we have the lira and you have the dollar' (Tagliabue 1985).

This book

This book is about how the seeds for the present 'American impasse' were sown many years ago in the origins and development of the United States in relation to the world-economy. But it is also about much more. It is about the peopling, geographical expansion, economic growth and political development of the United States in the world-economy. The impasse has been a long time coming and is, in one sense, only the latest in a long sequence of crises that have affected and afflicted the country. Each of these has had far-reaching effects on the United States as a whole and its internal geographical organization.

A primary theme of this book is the impact of the world-economy over time on the internal geography of the United States. While affecting the United States as a whole, the world-economy is viewed as having had impacts that have varied between regions. Indeed, regions have themselves often been defined by their relationships to the world-economy as much as, or more than, by their relationships to each other. For example, before Independence the various colonies had distinctive trading links with the mother country; today, foreign investment and the bulk of the new immigrants tend to cluster in distinct, and different, regions of the United States.

Attention turns first to the historical evolution of the United States within the world-economy (Chapter 2). Next the shifting regional impacts of the world-economy from the colonial period through until the recent past are traced (Chapter 3). Then the present-day impacts of the world-economy are examined with special attention to the regional geography of the United States (Chapter 4). A final substantive chapter (Chapter 5) briefly considers the major challenge facing the United States in the world-economy today: how to come to terms with the imperative of growth in the face of decline. A short Conclusion summarizes the major argument and the advantages of the book's approach.

2

The historical evolution of the United States in the world-economy: from incorporation to globalization

The involvement of the United States with the world-economy has followed a number of patterns of interaction. But these fall basically into two types. The first involves the *incorporation* of the United States *into* the world-economy and its development as a national economy. This covers the period from the original settlement by Europeans until the 'closing' of the frontier about 1890. The second involves the rise to dominance of the United States *within* the world-economy, the transformation of the world-economy this entailed in the form of the *globalization* of economic activities, and the problems this now poses for the United States. This period lasts from 1890 to the present. This division is somewhat arbitrary in that the seeds of globalization and its consequences were sown in the period of incorporation and national development. But 1890 does represent the time around which the United States began to emerge from its earlier 'peripheral' and 'semi-peripheral' status into a 'core' position within the world-economy.

The major framework used here for examining the historical evolution of the United States in relation to the world-economy is provided by Kondratieff's description of the world-economic 'long wave' (see Figure 2.1). American incorporation into the world-economy and national development lasted from an earlier logistic (the colonial period) through the first two cycles. During this long time-period the United States was still largely dependent upon the European core. But by means of political independence, territorial expansion, massive immigration and the resolution of internal cleavages to the advantage of an industrializing strategy over a resource-economy one (primarily as a result of the Civil War), the United States became a major force in itself. This period can be subdivided into two eras: that of mercan-

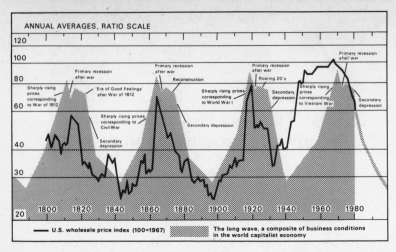

ANNUAL AVERAGES, RATIO SCALE

Figure 2.1: The relationship of Kondratieff's four cycles to American experience

tilism (1790–1840) and that of national industrialization (1840–90). During the second B-phase or global downturn (1875–96) the United States became a major threat to the previously hegemonic power, Britain, and also to aspiring powers such as France and Germany. So, progressively from 1770 onwards, the United States moved from semi-peripheral to core status. But there was nothing inevitable about this. Without the War of Independence and its success the United States may well have been peripheralized. Without the Civil War and its outcome the United States would probably have been unable to exploit its resources and other advantages within the world-economy to such great effect. There was, then, nothing 'determining' about the opportunity provided by the second Kondratieff downturn. The United States was literally in the 'right place at the right time'; its history had prepared it to take advantage of this opportunity.

Since 1890 the United States has been a dominant presence within the world-economy. After eclipsing Britain and Germany during the III A phase (see Table 1.1), the United States became the hegemonic power within the world-economy during the period 1920–40. This was followed by a period of consolidation from 1940 to 1967. It was during this period that the United States imposed its stamp on the world-economy and, as will be argued below, also laid the groundwork for the threats to its dominance that have been apparent since the late 1960s. A process of incorporation and national development in the early period (1790–1890) was followed, therefore, by a process of the globalization of the world-economy (1890–1967) largely stimulated by American influence. This chapter is devoted to describing these two processes.

Incorporation and national development

The colonial period

Britain versus France

The eighteenth century was an era of struggle between Britain and France for colonial supremacy. The two countries were in constant rivalry all over the world – in Africa, in India and in North America. In North America the British and French colonies had many features in common. They were settled at about the same time, beginning in the 1630s. They were located on the Atlantic seaboard and in the Caribbean islands. The native populations were relatively sparse and technically unsophisticated, so that the Europeans in these regions, unlike the Spanish in theirs to the south, could not live off native labor. Since the British and French found no precious metals, they had to support their settlement by agriculture, fishing, lumbering and fur-trading (Dorn 1940).

The British colonies fell into three groups: Virginia and its immediate neighbors, which produced mostly tobacco; New England, with its little groups of religiously Non-conformist settlements, which engaged in fishing, lumbering, the fur trade and commerce; and the British West Indies, by far the most profitable because of their sugar plantations (Mintz 1985). One important characteristic of the British colonies, taken as a whole, was their relatively large populations of Europeans, which were much larger than those of the French. A second characteristic was their relative political independence. All the colonies had governors, executive councils and judiciaries appointed from London. But nearly every colony also had an elective legislative assembly, usually in conflict with the appointed officials. The most common quarrel was over the insistence of London governments that all colonial products be sent to Britain in British ships.

The French settlements in North America, however, had a number of advantages over the British ones. Using the St Lawrence River Valley as their main base from 1608 onwards, the French took advantage of the waterway system for which it was the main outlet to push westward to Lake Superior and southward to the Ohio River. After opening up these regions numerous forts were planted along a route stretching from the St Lawrence to 'Louisiana,' the immense basin of the Mississippi claimed by the French after 1682. The British colonies on the Atlantic seaboard were effectively encircled in a great arc running from the Gulf of St Lawrence to the Gulf of Mexico.

By the early 1700s the French not only possessed a commanding geographical position in North America; they also had an advantage in their colonies' greater social discipline and political cohesion. There were no elective legislative bodies in the French colonies. There was a reproduction of essentially feudal social relationships such as those found in France. Altogether the governors of the French colonies were able to command obedience from their settlers without the resistance their counterparts often experienced in the British colonies (Priestly 1939).

Such, then, was the line-up of the British and French empires in North America. But the rivalry between them ended in a complete British triumph and helped set the stage for Britain's rise to hegemony within the world-economy in the nineteenth century. Why did the British defeat the French? One reason was that French governments were less interested in overseas possessions than in European hegemony. Another reason was that many more people emigrated from Britain to North America than from France. By 1688 there were 300,000 British settlers concentrated in the narrow strip along the Atlantic coast compared to only 20,000 French settlers in the huge area of Canada and the Mississippi Valley. This reflected both the refusal of French governments to allow the emigration of religious dissidents, who constituted a large component of the populations of the British New England colonies, and the commercialization of agriculture in England, which had forced large numbers of people off the land and made them available for emigration (Parry 1971). At the time of the American Revolution the population of the British colonies amounted to around two million, or a third of the population of the English-speaking world. The mass migration explains in part why Britain was victorious over France in 1763, and why the American Republic defeated Britain a scant two decades later.

But there were other causes of the British success. One was British industrial development. Britain's industrial growth in the period between 1550 and 1650 was surpassed only by its growth during the industrial revolution after 1760. This growth had various impacts overseas. More capital became available for colonial development, an important asset given the heavy initial expenditures required by colonies. Unlike the Spanish, the British and the French had no bullion and no native labor force that could be readily exploited. Entire communities of Europeans had to be transported and provided with seed, tools and equipment. London governments were generally more committed to the capital outlays this entailed than those in Paris were (Parry 1971). Britain was more entirely 'capitalist' in outlook by the mid-eighteenth century than France was, where 'pockets' of capitalism were surrounded by a hostile feudal hinterland (Fox 1971). Britain's industries also provided cheaper but more durable goods. British fur traders were therefore able to offer Indians cheaper and better items (blankets, kettles, firearms) in return for their pelts. Finally, there was a superior British shipbuilding industry. Along with the greater awareness of English rulers about the significance of sea power in maintaining overseas territories, this explains in large part the superiority of the British navy during the long series of Anglo-French wars from 1689 until 1763 (Dorn 1940).

The colonial–commercial competition between Britain and France gave rise to a series of four wars. The first three wars, those of 1689–97 (King William's War), 1701–13 (Queen Anne's War) and 1743–8 (King George's War), were not decisive in North America. The French enjoyed the support of most of the native inhabitants, partly because their missionaries were more active than the British and partly because the steady advance of British settlement beyond the Appalachian Mountains was a much greater threat than the scattered French outposts. With their Indian allies, the French constantly undermined British expansion. But this did not settle the basic question of who would control the Mississippi Valley. This question was settled definitively in the fourth war, that of 1756–63 (the French and Indian War), when a for once decisive British government under Pitt (the Elder) concentrated its resources on the navy and the colonies and, using its colonial manpower advantages, overcame the French one fort at a time until winning a climactic victory at Quebec, the heart of French Canada, in September 1759. The fall of Montreal in 1760 spelled the doom of the French colonial empire in North America. In the Peace of Paris in 1763 Britain received from France the whole of the St Lawrence Valley and all the territory east of the Mississippi.

When the Treaty of Paris was signed, the British politician Horace Walpole was said to have remarked: 'Burn your Greek and Roman books, histories of Little People.' This pompous remark points up the portentous implications of France's loss of North America (and India, at the same time). America north of the Rio Grande was to develop in the future as part of the English-speaking world.

Table 2.1a. *The value of each region's export trade with each overseas area as a percentage of that region's total exports*

	Great Britain and Ireland	Southern Europe	West Indies	Africa
New England	18	14	64	4
Middle Colonies	23	33	44	0
Upper South	83	9	8	0
Lower South	72	9	19	0
Total	58	14	27	1

Source: Walton and Shepherd 1979: 80.

Colonial trade

The life of the vast majority of the million and a half people in the British colonies at the mid-point of the eighteenth century was still dominated by the rhythms of agriculture. But now a sizable merchant class also thrived in half a dozen coastal cities, its trading operations bringing one of the sources of wealth that distinguished the eighteenth-century economy from that of the seventeenth. The great bulk of the accumulated wealth of America was derived either directly or indirectly from trade (Rossiter 1953). Though some manufacturing existed, its role in the accumulation of capital was not great. A merchant class of growing proportions was visible in Boston, New York, Philadelphia, Newport and Charleston by mid-century, its wealth based on trade. Even the rich planters of tidewater Virginia and the rice coast of South Carolina depended upon ships and the merchants who sold their tobacco and rice in the European markets. As colonial production rose and trade expanded, a business community emerged in the colonies, linking the provinces to one another and to the world-economy by lines of trade and identity of interest (Virtue 1936).

But even on the eve of the American Revolution, nearly 90 percent of the population made at least part of their living from subsistence agriculture (Walton and Shepherd 1979). The economically most successful colonies were those which could produce commercial crops to meet demands from Europe. As trade between the Old World and the New expanded, specialized forms of production, based on regional advantages and complementarity of trade developed (see Table 2.1). The most important product in the trans-Atlantic trade from the mainland colonies was tobacco, the staple of the upper South – the Chesapeake colonies of Maryland and Virginia. This constituted almost half of the total value of commodity exports from the mainland

Table 2.1b. *The value of each region's import trade with each overseas area as a percentage of that region's total imports*

	Great Britain and Ireland	Southern Europe	West Indies	Africa
New England	66	2	32	0
Middle Colonies	76	3	21	0
Upper South	89	1	10	0
Lower South	86	1	13	0
Total	79	2	20	0

Source: Walton and Shepherd 1979: 82.

British colonies in 1750 and remained the dominant export through-out the colonial period. Favored by suitable soil and climate and a protected natural harbor, the upper South was the major region of commercial agricultural activity in colonial America.

But other regions were also involved directly or indirectly with commercial agriculture serving European markets. In the lower South, especially in South Carolina, rice was a major cash crop, supplemented by the production of indigo (for dye) and naval stores (pitch, turpentine, tar, hemp, etc.). The region between the Potomac and the Hudson, including New York, New Jersey, Pennsylvania and Delaware, comprised the Middle Colonies. The land of this area was fertile and readily tilled and gave a comparative advantage to grain production. Known as the 'bread colonies', this region produced large quantities of wheat, corn, rye, oats and barley for export to Europe and to other colonies. Southern Europe and the West Indies were especially important destinations for Middle-Colonial goods. The specialization of the West Indies in sugar production, based on rapidly expanding demand in Europe, had important consequences for the Middle Colonies. It generated an expansive market for foodstuffs, livestock and other goods more easily supplied from North America than from Britain.

The region with the least-developed commercial agriculture was New England. Despite repeated attempts with numerous varieties of crops, New England failed to produce any crop with a large overseas demand. Poor soils and an inhospitable climate allowed production only for subsistence and local trade. However, New England *was* a major trading area. A most valuable export from New England was 'shipping services.' By the late colonial period these services, in combination with those of the Middle Colonies, were second only to tobacco exports in value (Shepherd and Walton 1972). But besides being a carrier of other colonies' products, New England was also an important center of the fur, lumber and fishing industries. Overall, though, these extractive industries were already less important in the colonial period than New England's development as a commercial center.

Because of the relative scarcity of labor and capital, the limited size of the market in the colonies, and restrictions imposed by the British Navigation Acts, manufacturing was relatively small-scale and most manufactures were imported from Britain. But there were exceptions. For example, New England had a thriving shipbuilding industry as early as 1660. This was so important that by the end of the colonial period one-quarter to one-third of *all* British-owned ships had been built in the American colonies. In this case the high costs of transporting raw materials, especially timber, made construction cheaper in New England, where suitable timber was at hand, than in England, where most timber for ship construction had to be imported from the Baltic region. But this situation was unusual. Most manufacturing in the colonies was confined to a limited production of iron, lumber-milling and raw-material processing (rum-distilling, homespun textiles, brick-making, etc.) (Walton and Shepherd 1979).

Population growth

As noted previously in the discussion of the advantages the British colonies came to have over the French, one of the most important features of British colonization was the high rate of population growth in the colonies. This rate has seldom been matched in any other place or at any other time in history. The rate was so high – the population doubled every 25 years – that the gloomy parson Thomas Malthus called it 'a rapidity of increase, probably without parallel in history' (quoted in Potter, J. 1965: 631). The general trends of population growth for each region are shown in Figure 2.2. The growth of population in New England is noticeably similar to that in the upper South. By way of contrast, the Middle Colonies and lower South, which developed later, display more rapid rates, which enabled them to catch up with the older colonies. In each case early growth is largely explained by immigration, with natural increase becoming increasingly important over time. This process was especially marked for Europeans, but also true to a degree for the African slaves imported to work in the plantations of the South (see Figure 2.3). As Walton and Shepherd (1979: 52) put it:

Plentiful high quality land and suitable climate acted like a powerful magnet

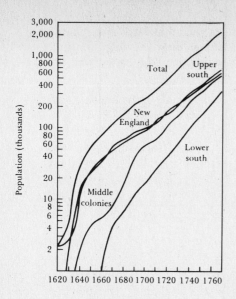

Figure 2.2: Population growth of the American colonies (Walton and Shepherd 1979: 51). Reproduced by permission of Cambridge University Press

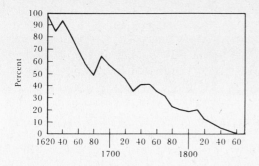

Figure 2.3: Foreign-born blacks as a percentage of the black population of the colonies and the United States (Fogel and Engerman 1974: 23). Reproduced by permission of Little, Brown and Company

to attract Europeans and capture and ship Africans. Also the fruits of the land sustained an extremely high rate of natural increase of the resident population. It was this combination of high rates of immigration and natural increase that forged this unusual record and gave North America its greatest crop, its people.

The European immigrants came mainly from England, though in the early eighteenth century there were substantial flows of Germans and the so-called Scotch–Irish (Scots previously resident in Ireland) to Pennsylvania and certain other colonies. Not including the substantial Puritan migration of 1630–40 not less than *half*, perhaps more, of all European immigrants were indentured servants, redemptioners or convicts (Hofstadter 1971). The tobacco economy of Virginia and Maryland was built upon the labor of gangs of indentured servants, who were replaced by slaves only during the course of the eighteenth century. Many of those indentured came to avoid destitution in an overpopulated England, but others were forcibly recruited. All bonded servants were chattels of their masters. Only the terminability of their contracts and certain limited legal rights separated them from slaves. After termination there was only limited social mobility. As Hofstadter (1971) argues: 'The Horatio Alger mythology has long since been torn to bits by students of American social mobility, and it will surprise no one to learn that the chance of emergence from indentured servitude to a position of wealth or renown was statistically negligible.'

The high rate of population growth in the colonies was partly determined by the high birth rate, which averaged 45 to 50 per 1,000 population as compared to 35 to 40 per 1,000 in England. Most migrants were young, married early and, because of the need for labor in agriculture, produced large families. But death rates were low, perhaps because of better nutrition and, in the absence of large urban populations, fewer epidemics. So altogether the rate of population growth was high. Of course this argument applies only to the Europeans. Even among them there were important differences. In particular, although European males could expect a long life of sixty years or so, European females faced higher mortality rates because of the hazards of serial childbirths. Their post-childhood life expectancy ranged only into the forties (Walton and Shepherd 1979).

Land and social structure
In America the availability of land rendered precarious those European institutions which were dependent upon the scarcity of land.

Efforts to establish feudal or manorial social systems largely came to nothing. The Dutch, who had settled parts of what later became New York, did try to establish a system of 'patroons,' or great landowners, whose lands were intended to be worked by tenants. But the patroonships did not last. Other attempts in Maryland and the Carolinas met with a similar fate. So in those areas where an attempt was made to introduce the 'traditional' social order of Europe, it failed to gain a hold. Quite early in the colonial period great disparities of wealth did appear, especially in the South, but this was stratification resting initially and finally upon wealth, not upon honorific or hereditary conceptions deriving from Europe (Main 1965).

This early system of stratification carried important implications for the future which transcend the matter of land tenure. In the first place it has meant that wealth rather than family or tradition would be the primary determinant of social stratification. Almost all Americans, regardless of class, have come to share a common ideology of utilitarianism or Lockean liberalism. Classes are therefore viewed as the outcome of differential *individual* efforts rather than social categories reproducing themselves over time. This has been possible in America because land or real estate has always been widely available, if only in small parcels, for private ownership. In the second place it has meant that there was never the basis for a conflict between aristocratic landowners and bourgeois capitalists such as developed in nineteenth-century Europe. All owners have been capitalists. Thus in America, and from the beginning, to paraphrase a President of the United States, the main business of America has been business.

Independence

It was remarked earlier that a notable characteristic of the thirteen British colonies was their large degre of political independence. The elective assemblies were frequently at odds with their governors and the other officials sent from London. It was also noted that Britain decisively defeated France in the French and Indian War and, by the Treaty of Paris in 1763, acquired France's colonies throughout North America. But once the threat from the French was removed a major incentive for the cooperation of colonists with the British was removed. Moreover, the British government decided, at what with hindsight appears to be a most inopportune moment, to tighten its imperial organization through a series of settlement restrictions and financial measures. The scene was now set for a major showdown between imperial authority and those with aspirations for colonial self-government.

Table 2.2. *Index of per capita tax burdens in 1765 (Britain = 100)*

Great Britain	100
Ireland	26
Massachusetts	4
Connecticut	2
New York	3
Pennsylvania	4
Maryland	4
Virginia	2

Source: Walton and Shepherd 1979: 163.

It is clear that not all, or even most, of the American colonists favored a violent rebellion against British authority. There were two major camps. The 'conservatives' wanted only to return to the loose relations between Britain and its American colonies that had prevailed before 1763. The 'radicals,' however, favored a complete break. Some of them also wanted a shift in political power inside the colonies in favor of popular participation. This was opposed by the conservatives, who wanted to retain the social status quo. They had no desire for democracy. Initially the radicals had their way – at least with respect to violent rebellion, largely thanks to the activities of British officials.

The steps leading to the American Revolution are well known and need not be related in detail. First there was the Proclamation of 1763 forbidding settlement west of the crest of the Appalachians. This was intended as a temporary measure to restrict population movement until a land policy could be formulated. But many prospective settlers and land promoters violently opposed even a temporary moratorium on 'opening up' the West. Then there was a series of financial measures – the Sugar Act, Quartering Act, Stamp Act and Townshend Duties – designed to shift a part of Britain's heavy tax load – empires are not cheap – to the colonists. These taxes and duties seemed reasonable to the British government since they had just spent a great deal of money defeating the French and argued that defending the frontier would involve considerable expense. Certainly, American tax burdens were relatively low (see Table 2.2). But the new levies were widely opposed by the colonists. A Continental Congress was called which organized a boycott of British goods until the financial measures were repealed. But another series of measures taken by the British govern-

because it created a new and *different* type of state that was later to transform the nature of the world-economy itself.

The era of mercantilism (1790–1840)

In the years immediately following Independence the advocates of federal authority were largely successful. In particular, the adoption of the Constitution in 1789 and the gradual emergence of a stronger central government laid the groundwork for the two most important features of early nineteenth-century America: the geographical expansion of the United States into the interior of North America and the development of American industrial capitalism. The shift from an Atlantic–commercial to a continental–industrial orientation symbolizes the political rejection of the peripheral status associated with the colonial period and the beginning of 'upward mobility' within the world-economy.

If the 'Hamiltonians' won in the sense that their vision of a commercial and industrial America was put into place, the 'Jeffersonians' did not completely lose. Their ideology of suspicion of government and big business established an *oppositional* stream in American politics that has provided a reservoir of ideals and rhetoric upon which numerous radical and democratic groups have drawn – Jacksonian Democracy, Populism, the New Deal, to name only a few. But in office, and beginning with Jefferson himself in 1800, the differences between Hamiltonians and Jeffersonians have been a matter of tone or style rather than substance. Jefferson's administration continued Hamiltonian policies, in particular the territorial expansion of the United States and the protection of domestic manufacturing.

Post-Independence trade

In the aftermath of the Revolution the colonists had been thrown back on their own devices and forced into local production and self-sufficiency. This had long-term consequences. Although trade recovered in the 1790s, a pattern of production for local use had been established. The Napoleonic Wars further stimulated American manufacturing by opening up both American markets and markets in the Caribbean and Latin America that had previously been dominated by the British. In the 1790s the United States took advantage of its new freedom to trade directly with countries in northern Europe and non-British colonies in the Caribbean from which it had been excluded in the colonial period by British mercantilism (see Table 2.3). Most of this trade was with France and the Netherlands, and the major com-

Table 2.3. *Average annual real exports to overseas areas: the thirteen colonies, 1768–72, and the United States, 1790–2 (thousands of pounds, 1768–72 prices)*

Destination[a]	1768–72	% of total	1790–2	% of total
Great Britain and Ireland	1,616	58	1,234	31
Northern Europe	—	—	643	16
Southern Europe	406	14	557	14
British Caribbean	759	27	402	10
Foreign Caribbean	—	—	956	24
Africa	21	1	42	1
Canadian colonies	—	—	60	2
Other	—	—	59	1
Total	2,802	100	3,953	100

[a] Northern Europe includes continental European countries north of Cape Finisterre. Southern Europe includes Spain, the Canary Islands, Portugal, Madeira, the Azores, the Cape Verde Islands, Gibraltar and other Mediterranean ports in Europe (except France). The foreign Caribbean includes Swedish, Danish, Dutch, French and Spanish colonies, Florida and Louisiana. Africa includes North Africa, the west coast of Africa and the Cape of Good Hope.
Source: Shepherd and Walton 1976: 406.

modity involved was tobacco. But Britain remained the most import-ant trading partner even after Independence.

One important change apart from the diversification of trading partners was the relative decline of the South, especially the lower South, as a trading region. The increase in exports between 1768–72 through the 1790s was due almost entirely to the increased exports of New England and the Middle Atlantic regions (Walton and Shepherd 1979). Especially important was the growing significance of New England's shipping and commercial services. This reflected the dis-appearance from the seas during the Napoleonic Wars of ships other than the British. The New Englanders took up the slack in the world trade, so to speak. But the 1790s also witnessed the invention of the cotton gin as a means of processing raw cotton and thereafter the lower South was to develop as a major supplier of cotton to the world market.

Territorial expansion

After Independence, therefore, America remained firmly a part of the world-economy in which it had originated. Both materially and ideo-logically it remained tied to Europe. If in the realm of revolutionary rhetoric there was an explicit rejection of European connections and alliances, there were in practice constraints upon a new state in a hostile world that led to easy victories for those elements in the American elite most anxious to participate in, and profit from, the European world-economy.

In 1800 the United States stretched only to the Mississippi River in the west. Though its northern border was much as today the country did not as yet extend to the Gulf of Mexico in the south. Florida, which then extended along the coast to the Mississippi, was still controlled by Spain. But within less than 55 years – that is, the lifetime of an early American European male – the United States was to attain its present size, an increase in area of some 300 percent.

It is hard to see how population pressure could have led to this vast territorial expansion. There was abundant land still available within the original land area. One factor was serendipity. The Louisiana Purchase, as is well known, was accidental. With a few pen strokes – and for only $15 million – Jefferson began a process of land acquisition by annexation and purchase that was to extend the United States to the Pacific in the west and into Mexico in the southwest (see Figure 2.4).

If Napoleon had not needed hard cash in 1803 *perhaps* no expansion, save perhaps in the vicinity of New Orleans, would have occurred! Of course, there was more to it than this. As Harriet Martineau, an English traveler, observed in the 1830s: 'The possession of land is the aim of all action, generally speaking, and the cure for all social ills among men in the United States. If a man is disappointed in politics or love, he goes and buys land. If he disgraces himself, he betakes himself to a lot in the West' (Martineau 1837: 91). But there was also the desire to open up the 'treasure chest' of natural resources for American industrial development. Very few of the new farmers were transplanted urbanites. Nearly all had farm backgrounds in the Northeast or in Europe. The incorporation of the vast trans-Mississippi region, and the lands acquired from Mexico by conquest in the 1840s, therefore provided both something of a social safety valve, as suggested by Martineau, and, more importantly, an immense resource base.

Frequently, the 'hunger' for land ran ahead of the boundaries of the state. The War of 1812 was largely stimulated by frontier settlers wishing to advance beyond their then limits into Canada and Florida (Pratt 1957). The Mexican War had similar origins in the 'revolt' of American settlers in Texas. All this was justified, of course, as the outcome of a providential course (see Chapter 1).

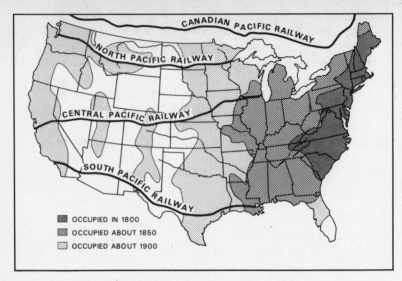

Figure 2.4a: The continental expansion of the United States: how the United States was settled (Chaliand and Rageau 1985: 77). Reproduced by permission of Harper and Row

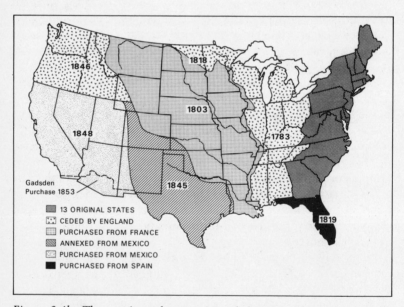

Figure 2.4b: The continental expansion of the United States: territorial growth (Chaliand and Rageau 1985: 77). Reproduced by permission of Harper and Row

The huge territory, however, presented problems. From a political standpoint, the new lands, even before they were settled, stimulated the increasing conflict between North and South. It was over the extension of slavery into the trans-Mississippi region that the issue of slavery finally boiled up into secession and civil war. From an economic point of view the problems of communication and transportation were immense. These also had important political implications. There was nothing inherent in the vast new country that would produce an integrated nation. Only the dramatic improvements in transportation brought about by canal construction, steamboats and railroads provided a basis for the economic and political integration of the United States.

Early industrial capitalism

The greatest phase of American territorial expansion coincided with the turndown of the global economy in the 1820s and 1830s. There was a particularly severe contraction in the English textile industry. This may have been the result of rising machinery costs causing the rate of profit to level off. More likely it was a 'realization' crisis in which low wages in England diminished domestic consumption while foreign markets seemed saturated. Whatever the cause, it triggered contraction across the board. But one consequence of vital significance to the United States, apart from the stimulus to more trans-Atlantic migration, was the rapid infusion of British capital into American railroad development as a means to resolve the profits crisis. Railroad-building in America, and in Britain, both restored capital accumulation and opened up new markets. It stimulated industrial development through its twin satellites of steel production and coal-mining. It also opened up the new lands of America to the world-economy.

State and local governments in the United States were important agents of this transformation. Some early 'model' projects in transportation engineering, such as New York's Erie Canal linking New York City to Buffalo and then to the West (1817–25), were entirely state-built. But the commonest mode of public assistance to transportation improvements was the so-called mixed enterprise, in which private and public funds were pooled. Apart from the real improvements brought about in internal communication, public assistance to economic enterprise in the construction of railroads and canals during the period 1820–40 had important long-term implications (Goodrich 1960; Chandler 1965). It suggests that government aids like land

grants to railroads after the Civil War, and even much later interventions such as the Tennessee Valley Authority, are very much within
a long American tradition of state-supported business enterprise
(Lively 1955).

By 1836 British investors held $200 million in American securities
largely connected with railroad construction. But in that year the
mania for railroads turned to panic in a complex financial crash that
led the British investors to abandon the American scene temporarily
(Jenks 1927). By the 1830s, then, there were complex trans-Atlantic
connections in trade, commodity prices and capital flows that made
each national economy subject to pressures and crises developing in
the others (Collman 1931). At this time the United States, rather than
generating such crises itself, was largely on the receiving end of
pressures from elsewhere, particularly Britain.

A concomitant of the extension of transportation and communication lines was the expansion of manufacturing; to a certain extent
the one stimulated the other. Before 1816 the so-called 'industrial
revolution' cannot be said to have come to America. In that year the
first major protective tariff was passed by the US Congress. Until then
there had been some manufacturing, as noted previously, but the
'factory' and the 'industrial worker' were still European, largely
British, phenomena. Certainly the Napoleonic Wars and the War of
1812 had stimulated domestic manufacturing because hostilities cut
off the usual imports from Britain. But with the end of the war with
Britain in 1815, American manufacturers were faced with the threat of
the renewed import of cheaper British goods. As a consequence a
powerful movement for a protective tariff was established. The
breadth of the movement is revealed by the fact that even old
Jeffersonians like President Madison supported it. Compared to later
tariffs such as those of the years after the Civil War, the Act of 1816
was modest indeed, but it represents the beginning of a significant
feature of nineteenth-century America: the governmental protection
of American industry against foreign competition.

By the 1830s factories were sufficiently widespread, especially in
New England and the Middle Atlantic states, to allow one to speak of
the onset of industrialization. The Northeast was favored by a history
of prior manufacturing and a wealthy merchant class. But wage costs
for skilled labor were considerably higher in the United States than in
England (Rosenberg, N. 1967). Consequently, at an early stage
American manufacturers resorted to capital-intensive methods of
production using cheap low-skilled labor. Continuous-process and
interchangeable-parts manufacture were the most widespread.

But until after 1840 other economic activities were relatively more important than manufacturing industry. One of these was shipping and the re-export trade. The Napoleonic Wars in particular stimulated the growth of an American merchant fleet, as noted earlier. This had various multiplier effects. With the income from shipping, American merchants and shippers increased their demands for certain subsidiary or complementary services such as brokerage, marine insurance, financing, warehousing and docking. Directly and indirectly, shipping and long-distance trade encouraged the urbanization of the Northeast. Between 1790 and 1810 the urban population grew from 5.1 percent to 7.3 percent of the total population, most of the increase occurring in the four port cities of New York, Philadelphia, Boston and Baltimore.

The other economic activity was the cotton trade. Especially after 1815 this was a major force in American development (Bruchey 1967). With income earned from cotton (and other staples) southern planters imported luxury goods, purchased the services necessary for marketing their exports, and bought much of the food necessary for sustaining themselves and their slaves. Cotton linked the United States to the world-economy directly, strengthening bonds with Britain in particular. But also by importing from the Northeast the services necessary for transporting, financing, insuring and marketing cotton, the South stimulated the further urbanization of the New England and Middle Atlantic states, which increasingly supplied the South with manufactures. Moreover, the South imported increasing amounts of food from the West and with the income received the West in turn bought manufactured goods from the Northeast (North 1961). Thus a regional specialization with between-region links was established that both 'integrated' the country and created three distinctive regional economies within it.

The era of national industrialization (1840–90)

If liberation from colonialism removed the political brakes from both geographical and economic expansion, the form of regional development that arose in consequence began to set its own limits. By 1840 the world-economy had 'bottomed out' and the next twenty years were to witness a major boom. But it was under these favorable circumstances that conflict between the two most distinctive regional economies of the United States was to reach its greatest level and bring about the Civil War. The Civil War did resolve the problem. It also led to the emergence of the United States as a major industrial economy as the

Table 2.4. *Manufacturing, by sections of the United States, Census of 1860*

	Number of establishments	Capital invested ($)	Employment		Annual value of products ($)	Value added by manufacture ($)
			Male	Female		
New England	20,671	257,477,783	262,834	129,002	468,599,287	223,076,180
Middle	53,287	435,061,964	432,424	113,819	802,338,392	358,211,423
Western	36,785	194,212,543	194,081	15,828	384,606,530	158,987,717
Southern	20,631	95,975,185	98,583	12,138	155,531,281	68,988,129
Pacific	8,777	23,380,334	50,137	67	71,229,989	42,746,363
Territories	282	3,747,906	2,290	43	3,556,197	2,246,772
Totals	140,433	$1,009,855,715	1,040,349	270,897	$1,885,861,676	$854,256,584

Source: Eighth Census of the United States: Manufactures.

Table 2.5. *US manufactures, 1860*

	Cost of material ($)	Number of employees	Value of product ($)	Value added by manufacture ($)	Rank by value added
Cotton goods	52,666,701	114,955	107,337,783	54,671,082	1
Lumber	51,358,400	75,595	104,928,342	53,569,942	2
Boots and shoes	42,728,174	123,026	91,889,298	49,161,124	3
Flour and meal	208,497,309	27,682	248,580,365	40,083,056	4
Men's clothing	44,149,752	114,800	80,830,555	36,680,803	5
Iron (cast forged, rolled and wrought)	37,486,056	48,975	73,175,332	35,689,276	6
Machinery	19,444,533	42,223	52,010,376	32,565,843	7
Woolen goods	35,652,701	40,597	60,685,190	25,032,489	8
Carriages, wagons and carts	11,898,282	37,102	35,552,842	23,654,560	9
Leather	44,520,737	22,679	67,306,452	22,785,715	10

Source: Eighth Census of the United States: Manufactures.

country adapted more successfully to the global downturn of 1870–90 than its European competitors.

The three regional economies

In the period 1840–60 manufacturing industry had spread well beyond its early centers in the Northeast. By 1860 Pennsylvania and New York both led Massachusetts in goods purchased and capital invested. The Census of 1860 revealed that over half a million men and women were working in some 53,000 factories in the Middle Atlantic states alone, and the value of manufactured goods produced in the United States was almost $2 billion (see Table 2.4). During the period 1810–60 the total value of products had increased from about $200 million to $2 billion. Capital invested had increased from $50 million to $1 billion (North 1961). The leading industry in 1860 was cotton manufacture, a New England industry (see Table 2.5). Lumbering was second, moving from its old centers in New England and the Middle Atlantic states to the West and the South. Of the first ten industries the milling of flour and meal was the only one in the West and South with a significant output. They were still 'peripheral' economies (North 1961). Iron manufactures, almost entirely a north-eastern industry, was not aggregated in the 1860 Census. Had all iron products and machinery been included in a single category they would have constituted the single largest category. Between 1850 and 1860 the doubling of the output of primary iron products and machinery forecast the shape of America's industrial future.

On the eve of the Civil War, therefore, an American urban working class was coming into existence. It would only be a short time before the dominance of the farmer in American society, if not in fable and politics, would be challenged by the urban worker. In fact the shift away from agriculture was well under way in the 1850s. In 1860 less than 60 percent of the labor force was employed in agriculture (12 percent was in manufacturing, the remainder in services and trade). Value added by manufacturing was still less than the value of the three major crops – corn, wheat and hay – and total capital investment in industry was still less than one-sixth the value of farm land and building (North 1961). But the movement away from farming and towards industry would accelerate after the Civil War. If in 1860 the United States was second only to Britain in manufacturing, it would within thirty years replace Britain as the industrial leader of the world (North 1961).

In the period from Independence until 1830 there was only limited immigration into the United States. But during the 1830s and the 1840s the influx increased, leading to the great movement that began about 1845 (see Figure 2.5). People from three countries constituted the overwhelming majority of newcomers in the years before the Civil

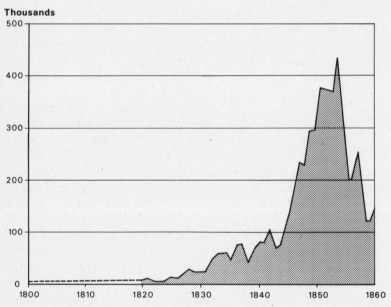

Figure 2.5: Immigration to the United States, 1800–60 (US Bureau of the Census 1957: 56–7)

War. A steady stream flowed from England until a decade after the Civil War. After 1830 Irish and Germans came in increasing numbers, pushed by economic and political conditions at home and attracted by the opportunities advertised in the 'new land.' The potato famine of 1845–7 precipitated a heavy Irish emigration that lasted well into the 1850s. The Census of 1850 reported nearly a million Irish in the United States, 40 percent of them in the large cities of the Northeast. The Germans came later following the failure of the Democratic Revolution of 1848. By 1860 1.3 million had arrived. Many Germans settled on farms in the Middle West, making use of the capital they brought with them. Others settled in the 'new' booming cities of the Middle West such as Cincinnati, Chicago, Milwaukee and St Louis. After the Civil War immigrants settled in increasingly large numbers in the larger cities.

The South shared neither in the new immigration nor in the upsurge of industrial activity before 1860. Of the 140,000 manufacturing establishments nationwide in 1860 only 21,000, or 15 percent, were in the southern states. Furthermore, southern production of manufactured commodities was only one-tenth of the national total at that time. This was to tell against the South once the Civil War was under way. But the South was not in fact an industrializing economy at all, the basis to its wealth and its social structure being out of step with the developments taking place elsewhere in the United States (see Chapter 3 for more detail on this). This is not to say that it was not a capitalist economy. To the contrary. But it was an *agrarian* capitalist economy based on slave labor and foreign export, rather than an industrial capitalist one based on wage labor and domestic industrial expansion.

In one sense, therefore, the American Civil War was the final act in the long struggle with British colonialism. The slave form of production in the South owed its continued existence and its prosperity to its integration into a British-dominated international trade. This blocked the integration of the United States under one mode of economic organization and threatened the frontier expansion. In particular, the tremendous industrial expansion in England after 1848 increased demand for agricultural raw materials, especially cotton. This encouraged the planters of the South to expand their domain, and slavery gained new footholds in the recently conquered lands of the Southwest, especially Texas. It thus threatened the policies and potential of northeastern industries and the interests of western farmers. The former required protection against British competition and access to the resource wealth of the West. The latter feared the extension of the slave system to the free lands of the West and were strongly attached to the doctrines of American uniqueness, robust individualism and equality at birth, which were tarnished by the continued existence of slavery.

For many years the different 'sectional' interests managed to coexist, if uneasily. At least until the 1830s southern cotton was important for northern economic growth in the form of exchange of trade and as a source of capital. But in the 1840s the pace of industrial growth in the Northeast accelerated to the point where the region became a manufacturing economy. This expansion ended the dependence of the American economy on a single agricultural staple. In turn this expansion created a demand for western agricultural products in the Northeast. Previously, agricultural surpluses from the West had tended to flow to the South to feed the specialized economy of that area. The 'transportation revolution' was vital to this transformation. The new railroads, in particular, stimulated the movement of farm products from the West to the Northeast. Western farmers were no longer dependent on the rivers of the Mississippi basin for marketing their produce.

The major consequence of all this was that by the 1850s southern planters could no longer look to the West for either allies or grudging support among farmers fearful of northeastern bankers and industrialists. Rather, by 1860 and secession, when the South sought to capitalize upon its control of the Mississippi, several thousand miles of railroad track between Chicago and the Northeast had weakened the pull of the Mississippi–Ohio transportation system. In that year over 19 million bushels of corn went to the Northeast as compared with only 4.8 million to the South; pork from the West to the Northeast was three times what it had been in 1850, while that of the South was only half the figure of ten years earlier. The railroads built during the decade of the 1850s thus helped to create the united front that the Northeast and the West presented to the South.

But ultimately, as Moore (1966) has cogently argued, the fundamental question over which the Civil War was fought concerned *which* economy – that of the South or that of the emerging North – should be favored by the machinery of the federal government. This was the meaning behind such matters as the tariff debates 'and what put passion behind the Southern claim that it was paying tribute to the North' (Moore 1966: 136). As Moore (1966: 136) continues:

The question of power at the center was also what made the issue of slavery in the territories a crucial one. Political leaders knew that the admission of a slave state or a free one would tip the balance one way or another. The fact that uncertainty was an inherent part of the situation due to unsettled and partly settled lands to the West greatly magnified the difficulties of reaching a compromise. It was more and more necessary for political leaders on both sides to be alert to any move or measure that might increase the advantages of the other. In this larger context, the thesis of an attempted Southern veto on Northern progress makes good sense as an important cause of the war.

The Civil War and its aftermath

1860 approximately marks the dividing line between a long swing of exuberant economic activity with generally rising prices and another long swing of fluctuating economic activity and falling prices, postponed somewhat by the war. Two of the major motors of American economic growth, railroad construction and cotton-textile manufacture, were severely restricted by the war. Workers suffered a drop in real income as prices rose by nearly 75 percent while wages rose by less than half that amount (Cochran 1961). And, of course, the South, to the extent that it contributed to American economic growth before the war, ceased contributing during the war and only slowly made its way into something like full participation afterwards. The slave-based economy was destroyed. But, in aggregate, the Civil War did stimulate American industrialization. After the first shock of disrupted trade, northern merchants and manufacturers prospered directly and indirectly in the business of provisioning and outfitting troops. After four years of war many kinds of industrial output increased rapidly. With this came increased profits and incomes for enterpreneurs willing and able to seize the opportunities offered. Many of the elite of late nineteenth-century American business made their fortunes in the Civil War (Salisbury 1962).

Historians like Charles Beard and Louis Hacker see in these wartime profits the *origins* of the Civil War. The northern business interest was bent on triumph by means of a 'Second American Revolution.' But the war was in fact stimulated by southern secession. The North was generally unprepared, as its performance during the war amply attests to. The evidence for the contention that the war was a 'revolution' is to be found *after* 1861 rather than before. To argue otherwise is a case of *post hoc, ergo propter hoc*.

What is the evidence? The evidence suggests that the Civil War produced a businessmen's government. Or, rather, government by, for and of businessmen. With the South gone from Congress, the business bias of the new Republican Party was quickly translated into legislation. The pre-eminence of an exclusively northern political party was proclaimed in a number of directions: in establishing under the

National Bank Acts of 1863 and 1864 centralized control over the banking and currency structure of the United States; in passing a very high protective tariff that in subsequent years was to prove to be one of the Republican Party's chief contributions to the rapid industrialization of the United States; in writing the Homestead Act as a stimulus to the private exploitation of the country's resources; in subsidizing railroad construction and railroad companies on a grand scale through land grants; and in encouraging the importation of 'cheaper' foreign workers to compete with domestic labor under the Contract Labor Law. Although some of this legislation was later modified or repealed, it nevertheless set the tone and content of the Republican Party platform for the next forty years. As a program it provided a climate in which the entrepreneurs of the 'gilded age' were to flourish mightily.

The industrial state

Who were the entrepreneurs and what did they do? One conclusion seems inescapable. They were not men who came up from the bottom. Most of them came from business or professional families (Mills 1945; Miller 1949). Something like 90 percent were native born. An Andrew Carnegie was an exception. So, as a group, the leaders of American business after the Civil War were not a 'new class' of men come to economic power. But their behavior and outlook were novel. Hence their characterization by some historians as 'robber barons.' Of 43 noted entrepreneurs studied by Destler (1946), 36 used monopolistic practices to gain their ends, 13 used political corruption, 16 milked their own companies for personal advantage, and 23 engaged in price-rigging. There is no way of knowing how typical this behavior was but there is no reason to believe that others were more scrupulous. What is important to note, however, is that American economic expansion in the era of national industrialization was built upon ruthless and often unethical business practices. Above all, 'the great game' involved using political connections and monopolistic practices to eliminate competitors. That was what late nineteenth-century entrepreneurship was all about (Josephson 1962).

During the 25 years that lay between the end of the Civil War and 1890 the American economy began to take on many of its modern characteristics. The change that stands above all others was the shift from an agricultural to an industrial economy. Until the 1880s agriculture remained the chief generator of incomes in the United States. But the Census of 1890 reported manufacturing output greater than farm output in dollar value. Soon, by 1900, the annual value of manufactures was to be more than twice that of agriculture. Relative to the rest of the world American gains in manufacturing output were phenomenal. In the mid-1890s the United States became the leading industrial power. By 1913 the United States was to account for fully one-third of the world's industrial production (Clark, V. S. 1929; Gallman and Howle 1971).

As industry expanded, railroads were extended across the country at a prodigious rate. In the first 30 years of railroad construction, from 1830 until 1860, some 30,000 miles of track had been laid. By 1873 the United States network was greater than that of all Europe, almost 200,000 miles. But fundamental to the revolution in production was the resource base of North America. The resources of the United States at that time can only be described in gargantuan terms: huge quantities of all the major minerals needed for industrialization – coal, iron ore, copper, minor metals like zinc, gold and silver, and huge quantities of oil; rich agricultural soils extending across thousands of square miles; a climate of sufficient diversity to permit long growing seasons for fiber crops like hemp and cotton and food crops like sugar; and rich forests providing vast amounts of lumber for housing and a wide range of wood products (Potter, D. M. 1965).

Superimposed on this foundation of plenty after the Civil War was a framework of political unity that encouraged the mobility of labor and capital. There were no linguistic, currency or customs barriers such as existed in Europe. America was a veritable empire at home, so to speak. Many of the raw materials and the markets Europeans had to obtain overseas were available ready at hand in the United States. But there was something more subtle, too. This was the attitude of sympathy with which government at all levels regarded business enterprise. Such a friendly atmosphere was most conducive to investment and entrepreneurship. For example, federal and state governments, not to mention municipalities, all seemed to feel that to aid the construction of railroads was to participate in the destiny of America. Foremost among the aids of government were the 134 million acres of public lands granted to some seventy *private* railroads by the federal government and the 48.9 million acres granted by eight states (see Figure 2.6). These gifts of land, if brought together, would comprise an area larger than Britain, Spain and Belgium combined. When not used for railroads they were sold off for private profit or kept for their mineral rights. It is inconceivable that private investors would have put their money into many railroads without the guarantees and gifts they received from all levels of government. The other side of the coin to the beneficence displayed by government

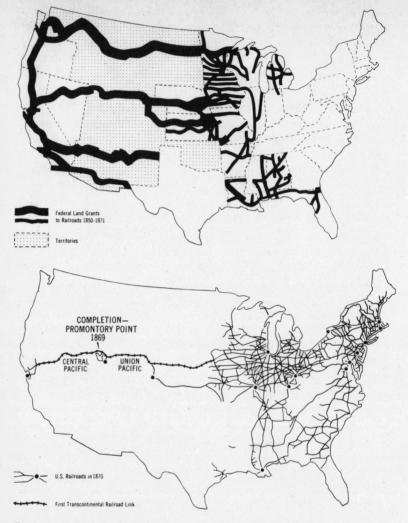

Figure 2.6: Railroads and land grants, 1870 (Robertson, R. 1973: 277).
Reproduced by permission of Harcourt Brace Jovanovich

toward business was the outright hostility displayed toward labor.
The massive federal intervention in the great railroad strike of 1877
and in the labor troubles at Coeur d'Alene, Pullman and Chicago later
in the century show this bias at its most overt.

The rapid industrialization was also encouraged by the great influx
of immigrants in the post-war era. Before the war, the annual immi-

Table 2.6. *American immigration, 1860–1920 (in thousands)*

Year	Number of immigrants	Year	Number of immigrants	Year	Number of immigrants
1860	154				
1861	92	1881	669	1901	488
1862	92	1882	789	1902	649
1863	176	1883	603	1903	857
1864	193	1884	519	1904	813
1865	248	1885	395	1905	1,026
1866	319	1886	334	1906	1,101
1867	316	1887	490	1907	1,285
1868	139	1888	547	1908	783
1869	353	1889	444	1909	752
1870	387	1890	455	1910	1,042
1871	321	1891	560	1911	879
1872	405	1892	580	1912	838
1873	460	1893	440	1913	1,198
1874	313	1894	286	1914	1,218
1875	227	1895	259	1915	327
1876	170	1896	343	1916	299
1877	142	1897	231	1917	295
1878	138	1898	229	1918	111
1879	178	1899	312	1919	141
1880	457	1900	449	1920	430

Source: US Bureau of the Census 1957: 56–7.

gration was measured in thousands but in the years after it it was often counted in millions. Between 1865 and 1880 more than 5 million newcomers found permanent or temporary homes in the United States. Between 1880 and 1920 the number swelled to 23.5 million. There were considerable year-to-year variations, the number of immigrants reflecting conditions both in Europe and in the United States (see Table 2.6). Peak years of inflow coincided with, or immediately preceded, the onset of severe depressions. Few Europeans could come without the immediate prospect of work. In times of rising economic activity and employment the tug on immigrants increased; as depressions came on and jobs disappeared the attractiveness of America diminished.

After 1880 the flow of people from northern and western Europe slowed and there was an increasing inflow from southern and eastern Europe. In the 1870s more than 80 percent of the immigrants came from the north and west of Europe; by 1910, 80 percent of the total came from the south and east. 1896 is often regarded as the year in which the switch took place. The new immigrants supplanted the old for two reasons: northern European labor was relatively better paid than earlier in the century and this reduced the attractiveness of emigration, and transportation changes had opened up eastern and Mediterranean Europe to trans-Atlantic travel.

American business profited greatly from the inexhaustible supply of skilled and unskilled workers. Manufacturing and mining companies profited most, for they could use the new labor to expand their operations without any increase in marginal labor costs. Moreover, the influx of immigrants meant more customers for American retailers, more buyers of cheap manufactured goods and an expanded market for housing. American businessmen were naturally convinced that unrestricted immigration was necessary for the growth of American industry. Labor organizations were equally certain that the influx of immigrants undermined the economic status of the already employed. Both were right and there was a continuous struggle until the 1920s about whether to restrict immigration or not.

But the wave of industrial expansion after the Civil War was a product of things as well as people. The hundreds of inventions of the period were in some cases of fundamental importance to the expansion of industry: the Siemens–Martin process in steel, which permitted even poor quality ores to be converted into steel, refrigeration, which laid the groundwork for a nationwide meat-packing industry, the telephone, the electric lamp, and so on. The list of such inventions would almost be a catalog of the industries which sprang up. In a country chronically short of skilled labor in its early years of industrialization, businessmen easily turned to machines to save on their wage bills. Many of the machines were not of American invention, but it was American businessmen who put them to their most intensive use. Coming late in the century, American industrialization was able to benefit from the improvements and inventions emanating from Europe.

European assistance to American industrialization amounted to more than technology. Thanks to the general political tranquility of the nineteenth century, the interchangeability of currencies afforded by the gold standard, and the attractiveness of American investments during profitability crises in Europe, American industry experienced little difficulty in attracting vast sums of capital from Britain and other European countries (Hall 1968). At any other time in history the process of American industrialization would have been much more like the strenuous ones of the Russians and the Chinese in the twentieth century. The advantages that accrued from tapping other people's money were tremendous. It has been estimated that Americans saved only between 11 and 14 per cent of their national income as against well over 20 percent saved by the British during their rise to eminence. Americans could thus both *spend* to stimulate domestic production and borrow abroad to finance capital investment, a double stimulus to industrial growth (Tarbell 1936).

American agriculture was also transformed by new technologies, business organization, and investment from abroad. But the period from 1865 to 1890 was one of great hardship for American farmers. Perhaps because of the ease with which land could be obtained, more people went into farming than demand for farm products was to warrant. All over the world fertile new lands were coming into production – in Canada, Australia, New Zealand and Argentina in particular. In the United States the number of farms nearly trebled between 1860 and 1900. Reinforcing this trend was an increased output per farmworker produced by mechanization and the application of new farming techniques (fertilization, new crop varieties, etc.). The basic problem was that agricultural productivity was too high relative to industrial production. Given that the supply of farm products increased rapidly after the Civil War, industrial output, with consequent increases in employment and incomes, would have had to increase even faster for agricultural prices to have been commensurate with farm costs (Frickey 1942). This situation changed after 1896 and ushered in a 'golden age' for American agriculture that lasted until World War I. But a pattern of market vulnerability was established as farming turned from being a way of life in its traditional connotation to being just another business. One measure of this was the increased indebtedness of farmers. Another was the increased reliance of individual farmers on giant meat-packing firms and food-processing companies for markets, transportation and finance.

The impact of the 'new' American agriculture in the world-economy was catastrophic, especially for many European countries. Between 1871 and 1891, for example, the price of wheat fell 27 percent in Germany and about 30 percent in Sweden. In the face of such declines many European farmers were ruined; many sought relief by moving to America, the source of their misery. This world-scale commercialization of American agriculture ensured a positive balance of trade for the United States throughout the period. But it also meant a loss of control. The price of wheat was now set by international forces. A poor harvest in America, which at one time might at least have had the mitigating consequence of higher prices, now meant nothing of the sort. A good crop in Russia or India could just as easily drive the price down, however well the American farmer farmed.

During the period 1865–90 commercialization led to an increased specialization among American farmers. With the creation of a national railroad net, eastern farmers were brought into competition with the richer farm lands of the West. The only recourse was either abandonment of farms, as in parts of New England, or switching to perishable crops for nearby urban markets (e.g. dairying, vegetables). In all cases the market was narrowing the freedom of choice available to the hitherto free-wheeling farmer. This led both to rural depopulation, especially when coupled with mechanization, and to the familiar geographical pattern of agricultural specialization that has been maintained in its major aspects to this day (see Figure 2.7). Competing in world markets put a premium on efficient organization and geographical specialization.

Railroad construction had pulled the US economy and the world-economy out of the slump of 1826–47, fueling renewed development in the years between 1847 and 1873 through a big spurt in industrial and agricultural production. The era of national industrialization was built upon the success of the railroad. But in 1873 expansion gave way to a downturn again. The very success of the railroad had fueled over-production and consequent lower prices. As yet, new technologies that could provide a new focus for investment were in primitive stages

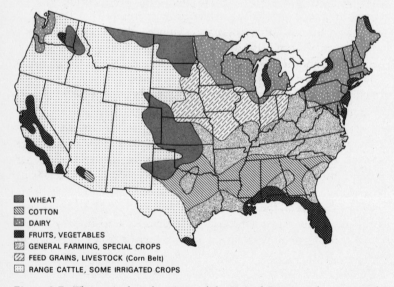

WHEAT
COTTON
DAIRY
FRUITS, VEGETABLES
GENERAL FARMING, SPECIAL CROPS
FEED GRAINS, LIVESTOCK (Corn Belt)
RANGE CATTLE, SOME IRRIGATED CROPS

Figure 2.7: The agricultural regions of the United States as they emerged in the 1880s (Robertson, R. 1973: 300). Reproduced by permission of Harcourt Brace Jovanovich

of development (e.g. the internal combustion engine and turbine, and the chemicals industry). However, it was in the United States and Germany that the new technologies were most developed; Britain, the steam power, lagged behind. So the industries of the United States were in a position to capture the new technologies for renewed expansion, having already almost caught up with Britain with the 'old' technologies. So 1873 not only signalled a general economic crisis; it also initiated a crisis in British hegemony. Thereafter Britain was in decline within the world-economy, all appearances to the contrary (Hobsbawm 1969). America was on the rise.

Globalization

The era of corporate growth (1890–1940)

This period has a number of distinguishing features. The first, revealed in the label given to the era, was the incredible growth in economic concentration through dramatic increases in the size of firms. The second was the emergence of the United States as a major actor in the world-economy. This meant both increased American economic involvement outside of North America and American participation in European-based power politics. The third feature, and one that became especially apparent at the end of the period, was massive intervention by the US federal government in the workings of the American economy. None of these features was entirely new. For example, government intervention in the economy was well established by the 1890s, and during the nineteenth century American relations with Latin America involved both territorial expansion at the expense of Mexico and economic expansion into Mexico, Central America and the Caribbean. Above all, economic concentration was under way in the 1870s. So in the period 1890–1940 these changes, rather than originating, came to fruition.

The setting of the 1890s

The great downswing in prices after the Civil War had meant continual trouble for the agricultural South and West. In both regions there was political upheaval as farmers fought for protection against the ravages of deflation. During the major depressions such as 1873–4 and 1893–6 there was also hardship in the industrial Northeast, where labor and small businessmen took the brunt of the downswing. The depression of 1893–6 was the culmination of more than twenty years' deflation. The physical suffering and psychological trauma that accompanied it were unprecedented and have only been equalled since by the Great Depression of the 1930s.

This depression occurred because a new engine of growth to replace the railroad had not yet been put in place and because the railroad itself had brought about changes in agriculture and manufacturing that produced a declining rate of profit. But other factors exacerbated it. One was government policies which encouraged economic concentration and monopolistic practices but also protected domestic industry from foreign competition. At the same time labor was ineffectively organized and subject to the constant fear of lay-offs. Increasing productivity, therefore, was reflected in falling unit costs of production and lower prices to consumers rather than proportionate increases in wage rates. Puzzlingly, economic success produced failure rather than more success.

One element in bringing about the depression of the 1890s, therefore, was economic concentration, which will be discussed later. Another was ineffective labor organization. A discussion of this is worth the digression from the other themes. There was a labor movement in the United States in the nineteenth century and, particularly in the 1880s, it seemed to be gaining in reach and power. A new American working class arose in the late nineteenth century as wage labor emerged as the definitive working-class experience. Growing even more rapidly than the general population, which almost doubled between 1870 and 1900, the industrial labor force had expanded to more than a third of the population by the end of the century. But it was divided ethnically and by skill-level and increasingly separated geographically into urban–industrial enclaves. The former discouraged labour organization while the latter encouraged it. However, low wages and miserable working conditions did stimulate some union organization and considerable strike activity.

The problem for labor was *staying* organized. No other labor movement has ever had to contend with the fragility so characteristic of American labor organizations (Perlman 1928). Of the two major national organizations of workers which developed in the late nineteenth century only the cautious, pragmatic American Federation of Labor survived into the twentieth century. The other, the Knights of Labor, less accommodating and more promising, disappeared.

One cause of the fragility was a lack of an integrated social consciousness on the part of American workers. The franchise was already available to the worker, who thus could separate the question of citizenship from the question of class. Labor unions were seen solely as economic rather than the political–economic organizations they were in Europe. Loyalty slackened during 'good times' and off the job. Another cause was immigration, which separated workers into ethnic–linguistic groups of limited communication. Finally, the weakness of labor organizations reflected the power of what can be called an ideology of growth or the 'growth culture.' This could also be called 'Americanism,' as it was some years ago by the American Socialist Leon Samson. This is the idea that opportunities for individual advancement *always* exist in America whatever one's present situation if only the quest for economic growth is unfettered. Although it is clear that mobility was limited, nevertheless many workers did acquire real estate and savings that allowed them to believe that they were 'making it' (Thernstrom 1964). Because of greater educational opportunities, others could believe that even if they did not 'get ahead' their children could. Once rooted in practice, the cult of success through growth often developed into a respect or admiration for businessmen. In such a context, attempts to portray employers and, more abstractly, capitalism, as the enemies of working men were of limited appeal and success. American labor unions have suffered, and suffer, the consequence (Davis 1980).

Economic concentration

A major feature of this era, as noted earlier, was the shift in firm size within manufacturing industry. Until the 1870s most firms were small and the creatures of individual businessmen or entrepreneurs. But beginning in the 1880s this changed. The first phase of economic concentration coincided with the downturn of 1873. Some of the impetus was technological. American industry discovered that both mass production and 'scientific' management, maximizing plant efficiency, required large amounts of capital. Large firms could supply this. But this was far from the whole story. More importantly, many firms had overexpanded in the years before 1873 to meet the demands of an expanding national market. With the realization that excess capacity was forcing prices below the costs of production small businessmen engaged in a fury of pooling and merging. The result was a massive consolidation and centralization in a wide range of industries, especially those in consumer goods (e.g. leather, whiskey, sugar, kerosene, biscuits, etc.). The most famous firm produced by this wave of concentration was the Standard Oil Company (kerosene production).

The severe depression of 1893–6 brought concentration almost to a standstill. The return of greater growth in late 1896 marked the beginning of a new and even more massive movement towards a concentration of industry. Over the years 1898–1905 more than 3,000 mergers took place. Urbanization provided a major part of the stimulus. The growth of cities stimulated the demand for both consumer and producer goods. To meet the challenge of supplying the new urban populations, new kinds of consumer-goods industries producing new products emerged. Firms in these industries formed large organizations, vertically integrated above the raw-material stage, in order to achieve economies of scale and eliminate competition. Among these industries were food processors, cigarettes and flour, as well as makers of sewing machines and typewriters.

An even more spectacular consequence of urban growth was the increased demand for producer goods like steel, copper, power machinery and concrete – the stuff out of which cities are built. This demand led to the formation of large, vertically integrated firms, extending from the mining of raw materials to the selling of finished products. A famous example of such a firm was the Carnegie Company, which by the early 1890s was an integrated firm owning vast coal and iron deposits. By 1901 and through a series of mergers the Carnegie Company was transformed into the world's largest corporation, United States Steel, with a capital stock of over one billion dollars, controlling 60 percent of the US steel business and owning a large part of US iron-ore reserves. While protecting its position in raw materials the Carnegie Company was now able to fix prices in an industry typified by high capital costs.

By 1905 roughly two-fifths of the manufacturing capital of the United States was controlled by 300 corporations with an aggregate capitalization of over $7 billion. Some of these companies were later broken up, for example Standard Oil and American Tobacco, and the pace of concentration slackened through the 1920s. But in the 1920s the process resumed, especially in the new automobile industry. More importantly, massive companies had become the reality of much American business.

America goes abroad

Besides urbanization another stimulus to economic concentration came from American expansion overseas. Bigger firms could better handle the initial costs and political chicanery involved in overseas expansion. Moreover, once big because of domestic stimulus, large firms were able to consider further growth only by expanding overseas (Wilkins 1970). The 1890s were particularly noteworthy because for the first time American governments became involved on a large scale in both facilitating and legitimizing American economic activities abroad. Lafeber (1963: 407–8) captures the process:

Sometimes the State Department seized the initiative in making the search [for markets], as in the Harrison administration. Frequently the business community pioneered in extending the interests of the United States into foreign areas, as in Mexico in the 1870s and in China in the 1890s. Regardless of which body led the expansionist movement, the result was the same: the growth of economic interests led to political entanglements and to increased military responsibilities.

In the late 1890s a more aggressive approach became apparent. If the 'outburst' of the 'new colonialism' on the part of European states in the 1880s was stimulated in part by the American industrial challenge, it also encouraged Americans to follow suit. Kiernan (1974: 111) argues:

All imperialism in history has been largely imitative; it has seldom or never grown spontaneously out of the needs or impulsions of a single state. Greece caught the contagion from Persia, Rome from Carthage, Islam from Byzantium, Holland and England from Spain and Portugal. European states themselves were now to a great extent rushing after one another in search of colonies because they saw their neighbors doing so.

But more particularly in the context of the 1880s and 1890s 'economic growth was now also economic struggle' (Landes 1969: 240). British industrial and financial domination was increasingly threatened by German and American competition. Yet all three countries faced depression in the early 1890s. American business had been able to expand without colonization, often, in Kiernan's phrase (1974: 124), 'employing the old empire-builders as managers.' Until 1898 there is evidence that large sections of American business remained opposed to territorial expansion overseas (Pratt 1936: 256–7). This was not opposition to economic expansion but rather concern at the expense of territorial annexation (Pratt 1936: 256; Lafeber 1963: 408). 'Doves' turned to 'hawks,' however, at the sight of the European states setting about the subjugation of China. Mark Hanna, *éminence grise* of the McKinley administration, identified the *limited* conception of colonial policy arrived at during the Spanish–American War. He desired 'a strong foothold in the Philippine Islands' for then 'we can and will take a large slice of the commerce of Asia. That is what we want. We are bound to share in the commerce of the Far East and it is better to strike for it while the iron is hot' (quoted in Lafeber 1963: 410–11). Areas were acquired 'not to fulfill a colonial policy, but to use these holdings as a means to acquire markets for the

glut of goods pouring out of highly mechanized factories and farms'
(Lafeber 1963: 408) (see Figure 2.8).

The translation from 'isolation' to 'involvement' was unsteady and
often politically divisive. Of course, the move towards empire by the
United States was not new at all. From its beginnings the United States
had been an empire, both in its size and in its tendency constantly to

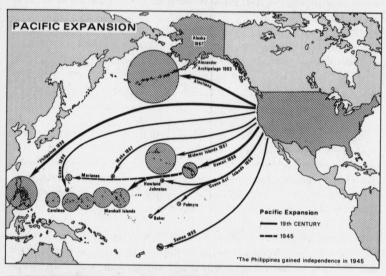

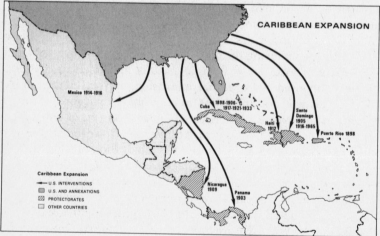

Figure 2.8: US territorial interventions and annexations after the Spanish–
American War (Chaliand and Rageau 1985: 79). Reproduced by permission
of Harper and Row

expand territorially. Nor was the expansion of the United States a peaceful operation. There were constant wars against the native Indian population, as Hollywood still reminds us, and a major war against neighboring Mexico in the 1840s. But most modern Americans – including historians – have accepted the view of contemporaries that the continental expansion of the United States was 'natural' and a part of 'manifest destiny.' To extend from ocean to ocean and from the Great Lakes to the Gulf seemed too geographically neat *not* to be part of a divine plan. What has seemed less natural, and therefore in need of more specific justification, has been the movement of the United States beyond its continental borders. This overseas expansionism or imperialism seemed to fly in the face of traditional American rhetoric about the evils of colonialism and the need for popular self-determination.

A major justification was that the rising recognition at home and abroad of the United States as a great industrial power caused many Americans to think about a new role for their country in the world. Americans took pride in their economic power and easily translated it into nationalism. It is, therefore, not surprising that during the 1890s a large number of patriotic societies, like the Daughters of the American Revolution and the Sons of the American Revolution, were founded. This was also the decade in which the cult of the American flag began. For the first time the flag salute and the flag itself were introduced into the public schools.

Some popular interest in looking outward clearly derived from the new national pride and the search for national prestige. But to some commentators it also reflected uncertainty and fear. Before the expansion in the 1890s there had been social unrest, hard times and farm discontent. The depression year 1893 witnessed huge strikes, marches of industrial workers upon Washington and the rise of widespread political discontent. Some conservatives understandably looked to war and international adventures as a means of meeting and redirecting unrest and discontent.

But there was a more mundane cause of why so many American political leaders and big businessmen were turning outward in their interests from the 1890s onwards, and this reorientation did not end with the Spanish–American War but continued with, if anything, increased intensity after the war was over. Quite simply, home markets were no longer enough for American manufacturing industry. But business expansion overseas did not necessarily entail territorial expansion. Indeed, colonialism in the European tradition was neither necessary nor desirable in most circumstances. What was important was that American business could have open access to foreign countries without the burden of everyday administration and the ideological stigma of colonialism. Pratt (1936: 256) reports that the *Commercial and Financial Chronicle* of 1898

believed that current European enthusiasm for colonies was based upon false premises; for although trade often followed the flag, 'the trade is not always with the home markets of the colonizer. England [Britain] and the United States are quite as apt to slip in with their wares under the very Custom-House pennant of the French and German dependency.'

This approach favored direct investment rather than portfolio investment and conventional trade. The best way of circumventing trade barriers was by setting up subsidiaries in foreign countries. This in turn favored larger firms. Advantages hitherto specific to the United States in terms of economic concentration – the cost-effectiveness of large plants, economies of process, product and market integration – were becoming the proprietary rights of firms. The world was now their oyster. Both American government policy and the needs of the larger US firms were thus conjoined. American governments could preach against colonialism while large American firms colonized the world.

During the first three-quarters of the nineteenth century capital exports were the main form of expatriate investment within the world-economy. But by 1910 a new type of investment activity came to dominate: the setting-up of foreign branches by firms operating from a 'home base' (Dunning 1983). In 1914 Britain was still by far the largest foreign capital stake holder, but the United States, though second in total, was much more committed to direct investment through subsidiaries (Dunning 1983: 86–8).

Much of US investment before 1914 was in other parts of the Americas. A particularly important sector of investment was tropical agriculture. American firms controlled sugar production in Cuba and banana production in Costa Rica, for example. But US manufacturing firms were most active in Europe, with 57.5 percent of US manufacturing subsidiaries located in Europe. Only 12.2 percent were located in non-industrial countries (Vaupel and Curhan 1974). There were therefore both supply-oriented and market-oriented multinational enterprises (MNEs) operating from the United States.

There was a deceleration in the expansion of US investment overseas through the 1930s. Nevertheless the US share of foreign capital stock, largely in the hands of MNEs, rose from 18.5 percent in 1914 to 27.7 percent in 1938 (Dunning 1983: 91). Much of this investment remained in the Americas but there were major new supply-oriented investments in South Africa, the Middle East, the Dutch East Indies and Liberia. Though the number of new subsidiaries continued to rise throughout the period, it was only in the late 1930s that the value of the direct capital stake equalled that of 1914. This was a period of consolidation and survival for US business overseas rather than one of unbridled expansion (Dunning 1983). This reflected the downturn in the world-economy in the 1920s and 1930s after the expansion of 1896–1914.

But the new American imperialism after 1896 was not simply economic. It was also political and cultural. Activist governments from McKinley through to Wilson encouraged American business abroad while also attempting to spread American culture and values. Of course, these were not unrelated activities. Spreading the 'American dream' involved encouraging the consumption of American manufactures – pushing the 'democracy of things,' in Boorstin's memorable phrase. But the package was bigger than this:

American traders would bring better products to greater numbers of people; American investors would assist in the development of native potentialities; American reformers – missionaries and philanthropists – would eradicate barbarous cultures and generate international understanding; American mass culture, bringing entertainment and information to the masses, would homogenize tasks and break down class and geographical barriers. A world open to the benevolence of American influences seemed a world on the path of progress. The three pillars – unrestricted trade and investment, free enterprise, and free flow of cultural exchange – became the intellectual rationale for American expansion. (Rosenberg, E. S. 1982: 37)

That the flow was all in one direction struck few as contradictory to the gospel of 'free exchange.'

The rise to power

In 1914 the United States was not immediately drawn into the war that had broken out in Europe as a result of political–economic competition between Britain and Germany. This reflected both a lingering American ideological isolationism and a sense that the war without American participation was probably more to American advantage than explicit American involvement would be. To President Wilson these 'principles' were expressed in the concept of 'neutral rights,' which asserted that the United States was free to trade in non-contraband goods with any belligerent it pleased and that its nationals were free to travel unmolested on belligerent ships. This was quite in keeping with American diplomatic tradition. As early as 1776 Madison and Jefferson had both emphasized the advantages to be gained by the United States as a neutral engaged in trade during a war. Jefferson put it best, if most cynically: 'The new world will fatten on the follies of the old.'

Eventually, of course, the United States did go to war and tipped the balance in favor of Britain and its allies against Germany and its allies. This resulted from an increasing perception on the part of American leaders that Germany represented much more of a threat than Britain to the US position in the world-economy. This was true both in terms of new technologies and businesses, where the German star had arisen alongside that of the United States, and in terms of 'style of operation,' where the Germans represented a continuation of European-style territorial imperialism in contradistinction to the American penchant for spreading the gospel of free enterprise, free trade and the American growth culture.

When Wilson led the United States into war he promised that a new international order would emerge from the struggle. It did. Not the international legal order that Wilson had in mind, but one based on American hegemony. America was the real winner of World War I. Economically Britain was crippled and became America's debtor. Germany, the country that had most threatened American aspirations, was defeated and exhausted. By 1920 the United States was 'a major creditor nation, and its bankers, merchants, shippers and manufacturers filled the vacuum left by the decline of the international monetary and trading system built by the British' (Becker and Wells 1984: 461).

Depression and federal intervention

Yet in only nine years the new American international order was faced with a critical test and the United States its most serious crisis since the Civil War. It came on suddenly, at least that is how it appeared at the time. In the pleasant summer of 1929 many Americans were congratulating themselves for having found a way to unending prosperity. The flow of goods and services had reached an all-time high, industrial production having risen 50 percent in a decade. A new 'mass-consumption' economy provided business with profits and workers with automobiles and household appliances. Farmers complained about price weaknesses in agricultural products, and there were real enough, but farmers were by the 1920s fewer in number and notorious for complaining. There seemed to be no reason to expect that production and prosperity would not go on increasing. The political

Table 2.7. *Estimated unemployment, 1929–43*

	Average annual number unemployed (in thousands)	Percentage of civilian labor force
1929	1,550	3.2
1930	4,530	8.7
1931	8,020	15.9
1932	12,060	23.6
1933	12,830	24.9
1934	11,340	21.7
1935	10,610	20.1
1936	9,030	16.9
1937	7,700	14.3
1938	10,390	19.0
1939	9,480	17.2
1940	8,120	14.6
1941	5,560	9.9
1942	2,660	4.7
1943	1,070	1.9

Source: Robertson, R. 1973: 682.

climate was favorable to business, then high in public esteem as the American 'provider' of material well-being. Reassuring news came from Europe. Socialism had been beaten back in most of Western Europe, war damage had been repaired, the gold standard had been restored and there was an increasing demand for American goods. Hope was high for a return to the free movement of goods and capital that had characterized the years from 1896 to 1914.

In the four years 1929–33 the American economy simply collapsed. The gross national product in current prices declined 46 percent, from $104.4 billion to $56 billion; in constant prices the decline was 31 percent. Industrial production fell by more than half; wholesale prices went down one-third and consumer prices one-quarter (Galbraith 1955). But the most horrifying statistics are those for employment and unemployment. Employment dropped by almost 20 percent and unemployment rose from 1.5 million to at least 13 million (see Table 2.7).

The profile of durable-goods production (automobiles, household appliances, etc.) from 1920 to 1945 reveals the magnitude of economic decline in the early 1930s (see Figure 2.9). From a peak in 1929 the index of durable-goods output fell to its lowest point in March 1933, the trough of the Great Depression. But unlike previous depressions when there had been a relatively quick rebound, this one seemed endless. Manufacturing output did not reach the 1929 level until late 1936 but then dropped off again until 1939. Only in August 1940 did production stabilize around the 1929 peak, more than eleven years after the start of the Depression.

These economic dimensions of the Great Depression are quickly sketched, too quickly perhaps to permit a full appreciation of the abyss into which America slid between 1929 and 1933. The jobless were everywhere. Starvation stalked the land. The ties of a family struck by unemployment were weak. Something like 200,000 young men and boys were reportedly traveling around the country in February 1933 for lack of anything better to do. What Robert and Helen Lynd (1937: 286) concluded for their study of Middletown (Muncie, Indiana) in 1935 can be applied to America as a whole:

The great knife of the depression had cut down impartially through the entire population tearing open the lives and hopes of rich as well as poor. The experience has been more nearly universal than any prolonged recent emotional experience in the city's history; it has approached in its elemental shock the primary experiences of birth and death.

Explaining the specific conjuncture of causes that brought about the Great Depression and fueled it for so long is no easy matter. But, as in the great downturn of 1873–96, overproduction under one technological regime without the timely appearance of a new one is a central part of most explanations. Schumpeter (1946), for example, finds the seeds of downturn before World War I in the mass production of consumer durables. The purchasing power of Americans (and others) had to increase at a high rate to keep up with increases in output. Consumer installment credit helped with this, at least temporarily. But in the 1920s earnings did not keep up with increasing productivity, so that by 1929 available markets were incapable of absorbing the full-employment output of American industry. In turn, as firms shed workers, workers were unable to keep up with credit payments. This produced shock waves through the financial system and in the stock market.

But there were other drags on the US economy in the late 1920s. One came from a dramatic downturn in the construction industry. A boom in the building cycle in the early 1920s had by 1925 turned to a bust as supply overreached demand. A second drag came from agriculture. During the 1920s the trend in world agricultural prices was downwards. This created a situation in farming regions remarkably similar to that of 1873–96; the more food that was produced the poorer farmers became.

One of the most striking features of the Depression years was the crisis of legitimacy that the American economy had to face. Driven by their desperation some Americans began to talk of violence and even of revolution. To Theodore Dreiser it seemed that Karl Marx's prediction 'that capitalism would eventually evolve into failure . . . has come true.' But many Americans, as the Lynds (1937) reported, were tempted by right-wing rather than left-wing responses. Order had to be reinstated, growth had to be restored.

The problem was there was no right or left to respond. To the extent . that there was a right it was represented by Hoover, the President when the Depression began and thus guilty by association. But Hoover was also an economic liberal rather than a right-wing authoritarian. The left was a mix of warring Socialists, Communists and Populists. Since the split in the Socialist Party in 1919 there was no movement on the left that provided a comprehensive critique of the existing society and a comprehensive vision of a new one.

What there was in 1930s America was a 'corporate liberalism' extending back to the turn of the century but coming to power under the banner of the 'New Deal' as the savior of America and its business

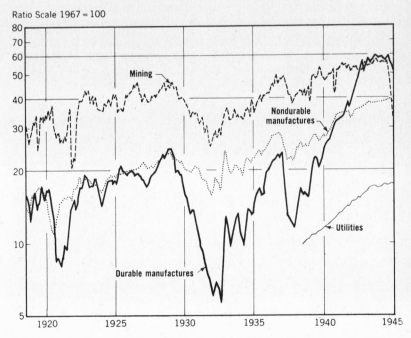

Figure 2.9: American industrial production, 1920–45 (Robertson, R. 1973: 698). Reproduced by permission of Harcourt Brace Jovanovich

civilization (Weinstein, J. 1968). Of course, business interests had enjoyed a predominant influence in American politics since at least the Civil War. But this relationship was relatively loose and did not go unchallenged. Beginning in the early years of the century, large firms increasingly opted for a policy of 'social responsibility' that went along with their domination of the US economy. Organizations such as the National Civic Federation tried to anticipate radical demands and sponsor reforms that would limit the growth of Socialism yet educate the business community to accept its responsibilities in stabilizing and legitimizing the system of large corporations that had just emerged. The Progressive Era (1900–20) is a monument to their success.

'Small' businessmen remained opposed to the growth of corporate liberalism. They were tied much more closely to market conditions than were many of the larger corporations. Their financial positions and profit margins were poorer. Their attitudes towards trade unions, working conditions and wages were correspondingly more rigid and uncompromising. The Presidents of the 1920s – Harding, Coolidge and Hoover – were their men. The coming of the Depression destroyed their political attempts to limit the tide of corporate liberalism. In 1932 it was the only alternative; the *laissez-faire* of the small businessmen was widely viewed as disaster.

So the New Deal was born out of a coalition of big business, labor and professional groups that had been in formation for many years. In 1933 it was finally lodged in Washington *in* the federal government.

As a governmental agent that recast American thinking on the responsibilities of government, however, the New Deal could only have been formed by an experience as jolting to Americans as the Depression. Although there is continuity between the reformism of the Progressive Era and the New Deal, wide differences in goals and accomplishments remain. The Progressive impulse was narrowly reformist: it placed some restrictions on business through regulation and inspection (meat inspection and the Food and Drug Administration originated then); it assisted agriculture; it freed labor from the extremes of *laissez-faire* capitalism, but it still conceived of the state as a policeman or judge and nothing more. It was also entirely a 'top-down' movement. For the New Deal under Roosevelt the primary and general innovation was the guaranteeing of a minimum standard of welfare for the American people through economic stabilization and social service policies. What is more there was a surge from below, a social-democratic impulse to it. The state was no longer an 'impartial' policeman. Few areas of American life were beyond the influence of the New Deal and its agents; even the once sacrosanct domains of prices and the valuation of money felt the tinkering. 'Tinkering with the system' became a standard feature of federal government activity with the New Deal.

Whether the New Deal actually 'ended' the Depression is a controversial question. The evidence suggests that it probably made some short-term difference. But the massive public spending of World War II was the ultimate solution (Robertson, R. 1973). What it did do was make life bearable or even possible for large numbers of people. It certainly headed off dissent. Above all it legitimized the idea of a strong federal government usually in partnership with, rather than opposed to, big business. This corporate liberalism was to be the guiding philosophy for the United States in the world-economy for the next thirty years.

The era of advanced corporate growth (1940–67)

From 1940 until the late 1960s the United States imposed its mark upon the world-economy by a mixture of bargaining, negotiation and bullying. As the new world policeman, replacing the now eclipsed British, the United States was able to impose policies of open trade in order to reap the rewards of its own efficiency. There were three specific aspects to this new role. The first is that the corporate liberalism of the New Deal laid the foundations for an aggressive partnership between American government and American business at home and overseas. A 'growth coalition' developed that dominated American economic and political life. Second, as a result of its hegemonic position, the United States was now the guarantor of the world-economy as a system of free trade and exchange. This involved new military responsibilities to keep challengers at bay and prevent defections from the capitalist world-economy into the camp of Communist autarkism led by the Soviet Union. Third, and finally, the United States replaced Britain as the kingpin of the world financial system. Wall Street in New York superseded the City in London as the world's pre-eminent financial center.

The new industrial state

The entrance of the United States into World War II, as we can now see, was a watershed in American and world history. World War I, it is true, had linked the political destinies of the United States and Europe, but that association had been brief and regretted by many Americans. By 1941, however, the events of the preceding years had

Table 2.8. *Major inventions, discoveries and innovations by country,*
1750–1950

	Total	Britain (%)	France (%)	Germany (%)	USA (%)	Others (%)
1750–75	30	46.7	16.7	3.3	10.0	23.3
1776–1800	68	42.6	32.4	5.9	13.2	5.9
1801–25	95	44.2	22.1	10.5	12.6	10.5
1826–50	129	28.7	22.5	17.8	22.5	8.5
1851–75	163	17.8	20.9	23.9	25.2	12.3
1876–1900	204	14.2	17.2	19.1	37.7	11.8
1901–14	87	16.1	8.0	17.2	46.0	12.7
1915–39	146	13.0	4.1	13.0	58.6	11.3
1940–50	34	2.9	0.0	6.7	82.4	8.8

Source: Dunning 1983: 106.

worked their effects. It was clear that the international order that the
United States had stood for after World War I had not been enforced.
In place of that order the Germans and the Japanese were forging a
world that would be unfriendly to the interests and ideology of the
United States. It became clear to many Americans that in this war they
could not stand aside for long. Finally the Japanese and the Germans
made the uncomfortable decision to go to war unnecessary: they
declared war on the United States.

It troubles many people to attribute material gains to war. The
suffering and loss of life associated with World War II should not,
however, blind us to the fact that this war lifted the American
economy out of stagnation. There was nothing 'artificial' about the
boom. The same effect could have been achieved, and was later
achieved in the 1950s and 1960s, by crisscrossing the country with
superhighways, and building hospitals, parking garages, schools and
vast residential suburbs (Gottdeiner 1985). Most importantly, the war
generated major technological advances. It speeded up the exploitation of discoveries from the 1930s (see Table 2.8).

The rapid increase in federal expenditures, which during the war
generated more than a third of the gross national product, provided an
experiment in curing depression and stabilizing the US economy
(Kuznets 1945). As a little more than half of total war expense was
financed by public (treasury) borrowing, the national debt increased
astronomically. But under wartime conditions this posed few problems, especially in relation to inflation. Specifically, because the debt
was internally held, Americans, either as individuals or banks,
acquired government securities to offset their liabilities. In addition,

because of rapid economic growth produced by deficit financing, individual and business assets were increasing faster than the rate of indebtedness. Finally, because the government could print money, the Federal Reserve banking system could always back up bank sales of securities.

The one threat was an unwillingness to tolerate higher taxes. This was met by imposing stringent price controls and rationing under conditions of nearly full employment. Thus higher incomes could not be used to speculate in goods and force up prices. Increasing taxes would have done the same thing. But World War II revealed a deep-seated resistance to income taxation on the part of many Americans as an unfair 'taking' by government from their earnings. After the war, and in the absence of price controls but with high government deficits, this was to prove a major impediment to the management of the US economy.

The war-time experience in 'curing' the Great Depression left the impression that government in coalition with business and labor – as during the war – could 'fine-tune' the US economy in such a way as to prevent the recurrence of massive depressions and maintain an ever expanding economy. Beginning with the Truman government, however, rather than continuing the *direct* intervention of the New Deal and war-time years, government policies were aimed at stimulating economic growth through indirect measures such as fiscal and monetary manipulation. Wolfe (1981a) has termed this 'Counter-Keynesianism,' even though it is usually called Keynesianism in the United States, because post-war macroeconomic policy in the United States, rather than using government to direct decisions made in the private sector, has used the private sector to influence the scope and activities of government.

The rise to dominance of the 'growth coalition' was a product of changes in the US economy long in the making. It all came together in the late 1940s as a result of World War II. Growth at home and expansion abroad were such clear possibilities that they unified the interests of previously contending groups in such a way that the nature of American political thinking underwent a major transformation in the 1950s.

As Goulden (1976) has argued in great detail, American political leaders in the late 1940s 'fell in love with' a 'Cold War plan.' Not only could American power be projected overseas but social change could be limited at home. This could be done by stimulating business and military activity overseas and by labeling domestic dissidents as friends of the Great Enemy (and recent ally) the Soviet Union. The

McCarthy period in American national politics was one in which anyone looking for social change in America was identified as 'subversive' or 'Communist.' This has had a legacy to this day. Much of America's faith in idealism – and politics being about anything other than growth – died in the early 1950s (Goulden 1976; Caute 1978). Wolfe (1981a: 28) sees it as follows:

When the New Deal was the decisive frame of reference, debate was divided into two camps called liberal and conservative. Liberals were those who believed that the government should play a positive role in correcting the abuses of capitalism by promoting a concern with equality and justice. Conservatives argued that business had made America great and that therefore as few reforms as possible should be passed that would undermine its privileges . . . From now on, a liberal was one who believed that growth should happen rapidly and a conservative, one who believed it should happen in a more tempered fashion.

Growth, therefore, became a substitute for choices concerning the distribution of power and wealth within the United States. Government policies rested on the premise that private groups (business) should be given public money to produce general growth. They used it to secure their power inside and, above all, outside the United States.

A good example of a government policy that was premised on stimulating growth through private enterprise was federal tax policy. Even a casual inspection of US tax history since 1945 shows that tax policy has been consistent in one area: the tax burden on big business has been eased by systematically shifting most of it to individual taxpayers (see Figure 2.10). This has been furthered to an even greater extent since 1980 (Edsall 1984: 212). The upshot of this has been to leave business with more funds for investment but at the same time to deprive the federal government of funds to cover its escalating costs, which include massive amounts spent on military protection for private business investments overseas!

The years 1940–67 witnessed a tremendous expansion of US business overseas. The United States accounted for about two-thirds of the increase in both capital stake and the number of new manufacturing subsidiaries in the period 1938–60 (Vaupel and Curhan 1974). In 1970 new investments in manufacturing, petroleum and mining abroad amounted to 35 percent of domestic investment in the same sectors (Calleo and Rowland 1973). A major trend was the selection of other industrial countries for new venture activities (see Table 2.9). Another was the shift from supply-oriented to market-oriented investments (Dunning 1983). Although this period saw the start of coerced divestment or nationalization of some primary-product investments and the setting-up of international producers' cartels, such as OPEC, it was only in the late 1960s that these problems for multinational business arose on a large scale. The rate of growth of the international capital stake was at that time impeded only slightly by these measures. But in the late 1960s the American share began to fall somewhat as businesses from other countries, especially Japan, became more active beyond their national borders (Dunning 1983) (see Table 2.10).

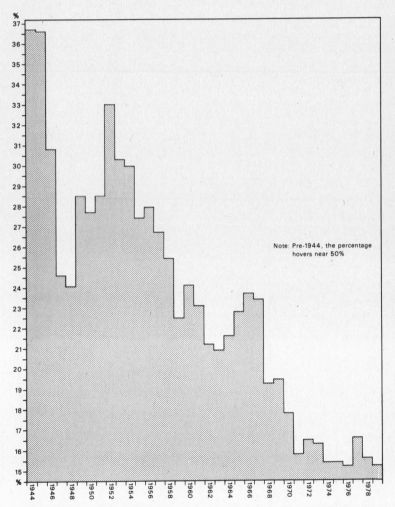

Note: Pre-1944, the percentage hovers near 50%

Figure 2.10: Corporation income tax as a percentage of IRS tax collections, 1944–79 (McDermott 1982: 146). Reproduced by permission of *The Nation*

Table 2.9. Estimated stock of accumulated foreign direct investment by recipient country or area, 1914–78

	1914		1938		1960		1971		1978	
	$m	%	$m	%	$bn	%	$bn	%	$bn	%
Developed countries	5,235	37.2	8,346	34.3	37.7	67.3	108.4	65.2	251.7	69.6
North America										
USA	1,450	10.3	1,800	7.4	7.6	13.9	13.9	8.4	42.4	11.7
Canada	800	5.7	2,296	9.4	12.9	23.7	27.9	16.8	43.2	11.9
Western Europe	1,100	7.8	1,800	7.4	12.5	22.9	47.4	28.5	136.2	37.7
Of which UK	(200)	(1.4)	(700)	(2.9)	(5.0)	(9.2)	(13.4)	(8.1)	(32.5)	(9.0)
Other European	1,400	9.9	400	1.6	neg.	neg.	neg.	neg.	neg.	neg.
Of which Russia	(1,000)	(7.1)	—	—	—	—	—	—	—	—
Australasia and South Africa	450	3.2	1,950	8.0	3.6	6.6	16.7	10.0	23.9	6.6
Japan	35	0.2	100	0.4	0.1	0.2	2.5	1.5	6.0	1.7
Developing countries	8,850	62.8	15,969	65.7	17.6	32.3	51.4	30.9	100.4	27.8
Latin America	4,600	32.7	7,481	30.8	8.5	15.6	29.6	17.8	52.5	14.5
Africa	900	6.4	1,799	7.4	3.0	5.5	8.8	5.3	11.1	3.1
Asia	2,950	20.9	6,068	25.0	4.1	7.5	7.8	4.7	25.2	7.0
Of which China	(1,100)	(7.8)	(1,400)	(5.8)	(neg.)	(neg.)	(neg.)	(neg.)	(neg.)	(neg.)
Of which India and Ceylon	(450)	(3.2)	(1,359)	(5.6)	(1.1)	(2.0)	(1.5)	(0.9)	(2.5)	(0.7)
Southern Europe					0.5	0.9	1.7	1.0	3.4	0.9
Middle East	400	2.8	621	2.6	1.5	2.8	3.5	2.1	8.2	2.3
International and unallocated	neg.	neg.	na	na	na	na	6.5	3.9	9.5	2.6
Total	14,085	100.0	24.315	100.0	54.3	100.0	166.3	100.0	361.6	100.0

Source: Dunning 1983: 88.

Table 2.10. *Estimated stock of accumulated foreign direct investment by country of origin, 1914–78*

	1914		1938		1960		1971		1978	
	$m	%	$m	%	$bn	%	$bn	%	$bn	%
Developed countries	14,302	100.0	26,350	100.0	66.0	99.0	168.1	97.7	380.3	96.8
North America										
USA	2,652	18.5	7,300	27.7	32.8	49.2	82.8	48.1	162.7	41.4
Canada	150	1.0	700	2.7	2.5	3.8	6.5	3.8	13.6	3.5
Western Europe										
UK	6,500	45.5	10,500	39.8	10.8	16.2	23.7	13.8	50.7	12.9
Germany	1,500	10.5	350	1.3	0.8	1.2	7.3	4.2	28.6	7.3
France	1,750	12.2	2,500	9.5	4.1	6.1	7.3	4.2	14.9	3.8
Belgium					1.3	1.9	2.4	1.4	5.4	1.4
Italy					1.1	1.6	3.0	1.7	5.4	1.4
Netherlands	1,250	8.7	3,500	13.3	7.0	10.5	13.8	8.0	28.4	7.2
Sweden					0.4	0.6	2.4	1.4	6.0	1.5
Switzerland					2.0	3.0	9.5	5.5	27.8	7.1
Others										
Russia	300	2.1	450	1.7	—	—	—	—	—	—
Japan	20	0.1	750	2.8	0.5	0.7	4.4	2.6	26.8	6.8
Australia										
New Zealand	180	1.3	300	1.1	1.5	2.2	2.5	1.4	4.8	1.2
South Africa										
Others	neg.	neg.	neg.	neg.	1.2	1.8	2.5	1.4	5.2	1.3
Developing countries	neg.	neg.	neg.	neg.	0.7	1.0	4.0	2.3	12.5	3.2
Total	14,302	100.0	26,350	100.0	66.7	100.0	172.1	100.0	392.8	100.0

Source: Dunning 1983: 87.

In the 1950s US firms moved abroad both to avoid trade barriers and to take advantage of labor cost differentials. The net effect of this, as Musgrave (1975) has shown, has been to the detriment of investment and employment in the United States, particularly because direct investment abroad involves the export of managerial skills and technology as well as capital. But in the 1950s, particularly as Western Europe and Japan were recovering from the dislocations of World War II, direct investment abroad was an attractive proposition for American business. Not only could uncompetitive markets be captured and risks spread, but processes increasingly expensive in American locations because of wage costs, taxes, etc. could be located overseas (Vernon 1966). The globalization of American business indicated that the United States was an increasingly unattractive place for the physical operations of much business activity.

American governments encouraged this overseas expansion by American business. American investment in Western Europe, for example, from the Marshall Plan onwards, was premised on a belief in the need to export the American model of business–government cooperation (Hogan 1985). The free movement of capital, free trade and anti-Communism combined as the major elements in American foreign policy. This worked for a time, until in the 1960s growth abroad, rather than reinforcing growth at home, began to undermine it. What was good for General Motors was no longer automatically good for the United States and, most certainly, vice versa.

Much of the growth in the US economy in the 1950s and 1960s depended upon increasing sales of consumer durables. This had also been the 'engine of growth' in the 1920s. The difference was that the 1950s was a time of expanding markets and relatively free trade for American business whereas the 1920s was the reverse. Another important growth sector was residential real estate. In 1900 only about one-third of non-farm households owned the dwellings in which they lived; the United States was definitely a nation of renters. By 1920 about two-fifths of households owned their homes. But during the 1940s and 1950s the number of owner-occupied units rose dramatically. By 1960 the percentage of home-owning households stood at 62.5 percent (Agnew 1981).

The rise in owner-occupancy and the concomitant rise in the consumption of consumer durables – they go together – were the result of three predominant causes. One major factor was the long period of steady income growth for many Americans. This encouraged both long-term 'investment' behavior and the assumption of long-term debts. However, there was not much *total* redistribution of income

Table 2.11. *Percentage of national personal income, before taxes, received by each income-tenth, 1910–59*

	Highest tenth	2nd	3rd	4th	5th	6th	7th	8th	Lowest 9th	tenth
1910	33.9	12.3	10.2	8.8	8.0	7.0	6.0	5.5	4.9	3.4
1918	34.5	12.9	9.6	8.7	7.7	7.2	6.9	5.7	4.4	2.4
1921	38.2	12.8	10.5	8.9	7.4	6.5	5.9	4.6	3.2	2.0
1929	39.0	12.3	9.8	9.0	7.9	6.5	5.5	4.6	3.6	1.8
1934	33.6	13.1	11.0	9.4	8.2	7.3	6.2	5.3	3.8	2.1
1937	34.4	14.1	11.1	10.1	8.5	7.2	6.0	4.4	2.6	1.0
1941	34.0	16.0	12.0	10.0	9.0	7.0	5.0	4.0	2.0	1.0
1945	29.0	16.0	13.0	11.0	9.0	7.0	6.0	5.0	3.0	1.0
1946	32.0	15.0	12.0	10.0	9.0	7.0	6.0	5.0	3.0	1.0
1947	33.5	14.8	11.7	9.9	8.5	7.1	5.8	4.4	3.1	1.2
1948	30.9	14.7	11.9	10.1	8.8	7.5	6.3	5.0	3.3	1.4
1949	29.8	15.5	12.5	10.6	9.1	7.7	6.2	4.7	3.1	0.8
1950	28.7	15.4	12.7	10.8	9.3	7.8	6.3	4.9	3.2	0.9
1951	30.9	15.0	12.3	10.6	8.9	7.6	6.3	4.7	2.9	0.8
1952	29.5	15.3	12.4	10.6	9.1	7.7	6.4	4.9	3.1	1.0
1953	31.4	14.8	11.9	10.3	8.9	7.6	6.2	4.7	3.0	1.2
1954	29.3	15.3	12.4	10.7	9.1	7.7	6.4	4.8	3.1	1.2
1955	29.7	15.7	12.7	10.8	9.1	7.7	6.1	4.5	2.7	1.0
1956	30.6	15.3	12.3	10.5	9.0	7.6	6.1	4.5	2.8	1.3
1957	29.4	15.5	12.7	10.8	9.2	7.7	6.1	4.5	2.9	1.3
1958	27.1	16.3	13.2	11.0	9.4	7.8	6.2	4.6	3.1	1.3
1959	28.9	15.8	12.7	10.7	9.2	7.8	6.3	4.6	2.9	1.1

Source: Kolko 1962.

Table 2.12. *Distribution of wealth by category, 1962*

	Wealthiest 20%	Top 5%	Top 1%
Total wealth	76%	50%	31%
Corporate stock	96%	83%	61%
Business and professions	89%	62%	39%
Homes	52%	19%	6%

Source: Ackerman *et al.* 1970.

and wealth then (see Tables 2.11 and 2.12), nor has there been since (Edsall 1984). But a major feature of the US economy in the 1950s and 1960s was an explosion of consumer debt (see Figure 2.11). This debt has to be serviced either internally or, as more recently, by the inflow of foreign capital attracted by high interest rates.

Another stimulus came from federal income-tax laws that allow mortgage interest as a personal deduction. Indeed, perhaps most significant in underwriting and securing growth through the domestic consumption of housing and consumer durables have been federally sponsored schemes which have greatly increased supplies of credit on easy terms. Urban mortgage debt increased ninefold from 1945 to 1962. Installment debt for the purchase of automobiles, furniture, household appliances, boats and recreational vehicles increased at similar rates, particularly in the 1950s.

This has been a classic example of post-war American 'Counter-Keynesianism.' Government has intervened in the economy to stimulate growth through the private sector and stimulate a culture of consumption that views more growth and more consumption as the twin measures of government success. Politicians compete by promising more. The American way of life became equated with the American 'standard of living.' The nineteenth-century 'producer ethic' based upon work, sacrifice, and saving evolved into a dominant twentieth-century 'consumer ethic.' But this consumer culture is more than a 'leisure ethic' or the 'American standard of living.' It is an ethic, a standard of living and a power structure. Life for most middle-class and many working-class Americans in the 1950s and 1960s involved seeking commodities as the key to personal welfare and even conceiving of themselves as commodities. One sells not only one's labor, but one's image and personality too. Meanwhile, the management of society is left to the few. The many are left to manage their consumption. Leadership by experts and pervasive consumerism developed symbiotically. The democratic promise of an economy subject to popular will was traded for increased personal affluence (see Fox and Lears 1983).

A major feature of the consumer culture is the importance of product differentiation. Successful differentiation requires that a product be widely advertised. Consumers must be persuaded that there is a good reason for buying one brand of cigarettes or cornflakes or washing machines rather than another. Advertising expenditures rose astronomically in the 1950s. This reflected both the fertile cultural ground that had developed and the needs of the large firms that both dominated production and could afford large outlays for advertising. In the 1960s, however, foreign, particularly Japanese, manufacturers were able to expand their sales in the United States by using the same strategy. Their strength was their remarkable success in persuading American consumers that their products were both of higher quality and greater durability than their American equivalents. The consumer culture thus exacted an economic as well as a political price.

Many Americans, particularly members of minority groups such as

blacks and Mexican Americans, never shared in the fruits of the consumer culture. Moreover, the retreat from the New Deal in the 1950s
left these people without support in a time of increasing *general*
prosperity. This relative deprivation set off a series of urban riots in
1964–6. A response to these, and to the problems of an aging population, came in the form of the Great Society programs of the Johnson
presidency – Medicare, the Office of Economic Opportunity,
Medicaid, various welfare devices, numerous housing programs
(Piven and Cloward 1972; Harrington 1976). These measures
increased federal government spending for social purposes tremendously over their previous levels, though they were still low by the
standards of other industrial countries (see Table 2.13).

A long-term consequence, however, has been the projection of
expenditures to maintain social harmony from an era in which they
seemed affordable into an era when they conflict with other expenditure needs (especially defense) in conditions of declining economic
growth. This is the so-called fiscal crisis of the American state
(O'Connor 1973).

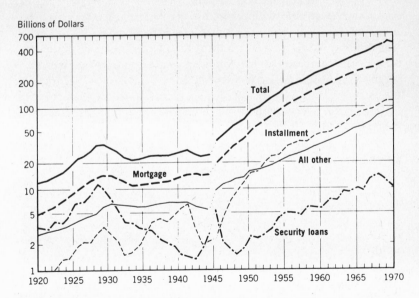

Figure 2.11: Kinds of debt in the United States, 1920–70 (Robertson, R.
1973: 645). Reproduced by permission of Harcourt Brace Jovanovich

The new American hegemony

American competition within the world-economy until World War II rested upon foreign investment, trade, technology and 'mass culture' rather than military might (Kiernan 1974; Rosenberg, E. S. 1982). But since 1945 military power has been added to the economic. At the end of World War II 'everyone else was crippled; an "American Century" seemed to be dawning' (Kiernan 1974: 126). The completeness of the victory over Nazi Germany and Imperial Japan was particularly important. It meant especially that Soviet influence flowed easily into German territory. When the war ended, Soviet armies were as far west as the River Elbe. This encouraged continuing American occupation of the territory American armies already controlled. But the Soviet Union also posed a long-term national and ideological threat to the United States. It was both a military competitor and the sponsor of an alternative image of world order. It had to be contained.

Containment is indeed the term that was coined to refer to America's policy towards the Soviet Union during the 1950s and 1960s. From the outset there were two prongs to containment policy. One was simply military – that is, to oppose by force Communist or Soviet military forces, or ones that could be construed as such, wherever and whenever possible. Examples would include Korea and Indo-China. The second was to make life sufficiently rewarding in countries threatened by 'Communist takeover' that they would be drawn towards the American model rather than the Soviet one. The best example of this strategy would be the Marshall Plan, by which the United States undertook in 1948 to rehabilitate Western Europe. This set the stage for the development of the European Common Market in the late 1950s, but in the interim also stimulated the growth of a formal military alliance between some West European countries and the United States (NATO).

Containment policy represented in practice the attempt to create a *Pax Americana*. The United States was to be the defender of a liberal international order, the 'free world' as it was called in the United States, if not elsewhere. The logic behind this lay in a presumed identity between the United States and world economies. Overseas expansion would stimulate economic prosperity abroad, thereby rebounding to American advantage. Military expenditures abroad would create a protective apparatus for American investment in other countries. 'Power projection' was viewed as a *legitimate* extension of commercial activities abroad (Keohane 1984; Cox 1981).

But as Louis Hartz (1964: 118) has pointed out, 'From the time of Wilson, indeed even before then, if we take into account a stream of

Table 2.13. *Public income and expenditure (central government, local authority and social security) in the United States and Western Europe*

| | Total receipts as % of GNP 1969 | Current expenditures[a] as % of GNP | | Investment[b] as % | |
		1959	1969	of GNP 1969	of national investment 1969
Sweden	49	28	37	6.5	29
Norway	43	22	37	7.0	28
Netherlands	42	27	37	6.0	28
United Kingdom	39	29	33	5.0	24
France	38	31	33	5.0	20
West Germany	38	30	32	4.5	20
Austria	38	27	31	4.5	19
Denmark	37	22	32	4.5	18
Finland	36	23	28	4.0	18
Belgium	34	28	33	4.0	16
Italy	33	27	33	3.0	13
Ireland	31	23	26	2.0	11
Switzerland	28	20	23	2.5	10
Greece	27	18	24	—	—
Spain	22	13	18	—	—
United States	32	25	29	3.0	17

[a] Current expenditure equals total expenditure minus investment and capital transfers.
[b] Gross fixed asset formation.
Source: Calleo and Rowland 1973: 178.

thought which accompanied our early imperial episodes at the turn of the century, the country has actually sought to project its ethos abroad.' The tendency of Americans to project their values upon others was dramatized in the view held at the highest levels of decision that the Vietnam War was a battle between two fixed groups of people with different but *negotiable* interests. That the others in this case were engaged in a social revolution never really came to mind. Ultimately, the persuasiveness of 'American ways' would be victorious. The continuity in American expansion is clearly expressed in the analogies to early America used by American combatants in Vietnam:

In Vietnam American officers liked to call the area outside GVN [Government of Vietnam] control 'Indian country.' It was a joke, of course, no more than a figure of speech, but it put the Vietnam War into a definite historical and mythological perspective: the Americans were once again embarked upon a heroic and (for themselves) almost painless conquest of an inferior race. To the American settlers the defeat of the Indian had seemed not just a nationalist victory, but an achievement made in the name of humanity – the triumph of light over darkness, of good over evil, and of civilization over brutish nature. Quite unconsciously, the American officers and officials used a similar language to describe their war against the NLF [National Liberation Front]. (Fitzgerald 1972: 491–2)

American expansiveness, therefore, owes as much to Main Street – the attitudes of the general population – as it does to Wall Street.

In the 1950s and 1960s global military commitments led to the creation in the United States of what President Eisenhower called a 'military–industrial complex.' Alone amongst post-war American Presidents, Eisenhower, the ex-general, was skeptical of overseas military expenditures and military spending in general (Wolfe 1981a). The others have believed that a 'war economy' was not only necessary for the *Pax Americana* but could be an instrument for domestic economic growth (Mintz and Hicks 1984).

By every available measure the war economy of the United States is a colossus. In the allocation of federal expenditures, payment for past, present and future military spending dominates. It was 48.6 percent of the federal budget in 1981 (DeGrasse 1984). The military element of government purchases in 1971 was 73 percent of the $93 billion of total purchases within the federal budget. As purchases of military equipment and fixed installations accumulated, they reached the value of $214 billion in 1970. In that year the total assets of all US manufacturing corporations was $554 billion. Hence the assets of the US military were 38 percent as much as the assets of all US industry (Melman 1974).

Of course, these assets deprived other sectors of the economy of fresh investment (Mosley 1985). Rather than stimulating the economy, military spending has had a retarding effect. Russett (1970), for example, shows that for each dollar spent on the military over the period 1939–68 there was $0.163 *less* expenditure for consumer durable goods, $0.110 *less* for producer durable goods, and $0.114 *less* for homes. Another way of examining the 'stimulative' effects of military spending is to compare countries with different levels of spending to see if there is any correlation with economic performance (investment level or labor productivity). DeGrasse (1984) suggests that the United States, with high military spending, was outperformed by other countries in direct relation to the level of military spending, that is the lower the military spending the better the economic performance (see Figure 2.12). Even during a period of massive economic growth, therefore, 'guns' took away from 'butter.' One result is clear: the lack of investment in plant and equipment in many US factories is no mystery. As Melman (1974: 67) puts it: 'US policy traded off renewal of the main productive assets of the economy for the operation of the military system.'

In the relations between the United States and other countries, military spending also figures highly. Fully half the US defense budget in the 1960s was tied to NATO. Though the West European states had

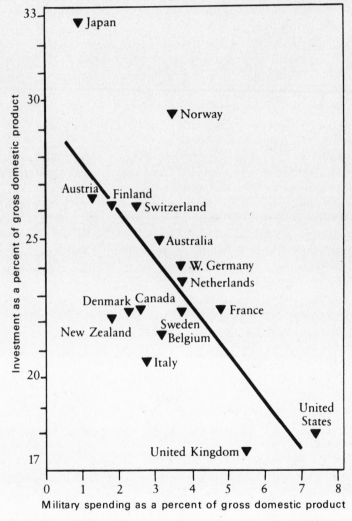

Figure 2.12: Investment versus military spending, selected nations, 1960–80 (DeGrasse 1984). Reproduced by permission of the *Bulletin of the Atomic Scientists*

then recovered to the point where their collective resources rivaled those of the United States, the United States continued to assume responsibility for a large proportion of Europe's 'defense costs.' This had become an enormous drain on the US economy. This is clear when one examines the US balance of payments in the 1950s and 1960s. Except for 1959 the US current account was always in surplus in the 1950s. But five large items threw the basic balance into deficit over time: overseas military expenditures, foreign aid, direct capital investments abroad by American corporations, short-term capital flows, and tourist expenditures. Most of these, except the short-term capital and tourists perhaps, were the costs of *Pax Americana*. They represented in Calleo's (1982: 21) words: 'official military or economic subventions or private capital investments to shape a new world economy.'

Global financial dominance

In the post-war era America's role as a military guarantor of the world-economy has been paralleled by its role as the kingpin of the financial system. Hegemonic monetary systems are not novel in history. But under the old gold standard system that prevailed in the nineteenth and early twentieth centuries, Britain operated more as a manager than as a tyrant. A liberal international order required this; the system needed defending rather than exploiting. This is not to say that benefits did not flow to the British. However, the British were an ever open market for other people's surpluses, a lender of the last resort in financial crises, and the source of long-term countercyclical lending. Ultimately, even this system broke down in 1931 because the needs of the system, for a reserve currency and lender of last resort, conflicted with the needs of Britain itself to reduce its money supply and its debts to foreign banks (Calleo and Strange 1984).

During the period from the Bretton Woods Agreement in 1944 until 1971 the United States was the reserve-currency country for the world-economy. Throughout this period and because of its general hegemonic position within the world-economy the United States was able to finance its deficits, the costs of *Pax Americana*, through its role as monetary overseer. The United States did not run deficits throughout the 1960s, then, to provide loans to others but as a by-product of pursuing domestic growth and overseas expansion at the same time (see Table 2.14). The Vietnam War was a particular drain.

But the Bretton Woods system of fixed exchange rates pegged to the dollar was a major boost to takeovers of foreign industries by American business. Under this system each nation was responsible for

Table 2.14. *US basic balance of payments and net military transactions compared ($ million), 1960–70*

	US basic balance	US net military transactions
1960	−1,155	−2,752
1961	20	−2,596
1962	−979	−2,449
1963	−1,262	−2,304
1964	28	−2,133
1965	1,814	−2,122
1966	−1,614	−2,935
1967	−3,196	−3,138
1968	−1,349	−3,140
1969	−2,879	−3,341
1970	−3,038	−3,371

Source: Calleo and Strange 1984: 104.

keeping the value of its currency within 1 percent of its 'par' or set value. To keep within that range, central banks (the Bank of England, the Bank of Italy, etc.) were required to sell or buy their own currency on foreign-exchange markets. By running large deficits the United States effectively forced foreign central banks to buy excess dollars with their own currencies to decrease the supply of dollars in global circulation. This provided American investors with the foreign currencies (francs, lire, pounds, etc.) necessary to buy assets in France, Italy, Britain and elsewhere. At the price of international 'monetary stability,' foreign central banks were put in the position of financing the takeovers of their own industries (Fusfield 1968).

In this financial setting American corporations were able to make massive foreign investments in new plant, producing commodities for foreign markets and importing back into the United States. During the 1960s direct shifts of US capital abroad were enormous. One corporation alone, General Electric, increased its overseas capacity by 400 percent – from 21 foreign plants in 1949 to 82 in 1969 (Babson 1973). By the early 1970s nearly one-third of annual US automobile company investment was abroad (Musgrave 1975).

This exploitation of the international monetary system could not continue, however, once the money supply in the United States exploded beyond the capacity of the world-economy to absorb it. America could no longer simply export the inflation generated by the US economy and the US government by devaluing other currencies.

The only reason that new external restraints on American monetary policy towards the rest of the world have not been forthcoming is that no other country and currency has as yet emerged to replace the United States and the dollar (Guerrari and Padoan 1986). This is why even after the collapse of the American-dominated system in 1971 the United States can still use monetary policy to its advantage (Parboni 1985).

The international monetary system, however, has become increasingly international (Makler *et al.* 1982). Because of new communication systems, satellites, computers, etc., and the shift from complete US control without a new 'national' replacement, capital markets throughout the world have become rapidly integrated. International banking has seemingly won independence from national constraints. This is reflected both in the tremendous increase in the international activities of American banks and the 'debt crisis' on a global scale consequent on their overexpansion of credit in the 1970s. The growing importance of international transactions has in turn increased the significance of international organizations set up to monitor the global financial system. These organizations – the International Monetary Fund, the International Bank for Reconstruction and Development (World Bank) and the Bank for International Settlements – have become well known since the early 1970s. The United States has tried to use these organizations to substitute for its loss of monetary hegemony (Sampson 1981). The problem is that unless the United States can put its own economic house in order its role within these organizations is likely to be a shrinking one. On all fronts, then – economic, political and financial – the globalization of the world-economy that American business and government have largely sponsored and sustained has come back to haunt its creators.

Conclusion

The argument of this chapter is that the history of the United States in relation to the world-economy has gone through two major stages: first, incorporation and national development and, second, globalization. During the first, covering the colonial period through the first two Kondratieff cycles of the nineteenth century, the United States was largely dependent. But during the second, covering the period from 1890 through until the recent past, the United States rose to a dominant position. The United States achieved and maintained this position without a territorial empire. This, however, opened the door to a process of globalization which, although beneficial to the United

States for many years, has recently become more problematic. After acquiring a territorial base at the continental scale, the American state tended to leave the business of expansion to business except when military enforcement became necessary. The Founding Fathers' fear of a strong state, the result of their own experience with eighteenth-century Britain, led to the easy usurpation of public power by private interests. But this has not been without advantages to the United States within the world-economy. The power and affluence of Americans and the United States in the twentieth century owe much to the weakness of the American state in the nineteenth.

Today, however, the national economy is no longer the basic building block of the world-economy, but has a rival in the immediately global market which can be supplied directly by firms capable of organizing their production and distribution without reference to state boundaries (Fröbel *et al.* 1977). This is not to say that the global will soon completely transcend the national. Transnational activities must still operate within states and under constraints imposed by them. But, compared to the past, national and international economies no longer reinforce one another. For political reasons, national economies cannot be confined solely to creating optimal conditions for the operations of business within them (MacLaughlin and Agnew 1986). At the same time, the global economy is developing explicitly to optimize conditions for business, at whatever cost to this or that national economy, including the firms' own.

The conflict between the national and the global is particularly acute in the American case. Giant multinational corporations have been the principal engines of saving and investment within the American economy. Their investments, however, are dictated by their own need to make increasing sales and profits. This had led to both overinvestment in manufacturing for consumption and, ultimately, because of market saturation, to vast investments abroad. The availability of capital for overseas investment contrasts sharply with its scarcity for private investment and public purposes at home. The exception, of course, is military spending. In this case American taxpayers must foot the bill for protecting those disinvesting from the United States!

Military spending has become a major drain on the US economy. Yet while it pretends to global hegemony the American state must maintain it. Today, therefore, the United States and its people are caught on the horns of a dilemma. On the one hand, recapturing economic growth requires a reduction of military expenditures. On the other hand, a reduction in military expenditures makes the world less safe for the multinational business upon which American growth was based in the past. This also brings into question American political and cultural hegemony. The dilemma requires a radical solution. 'History,' as Calleo (1982: 195) observes, 'is finally presenting America with its bill.' It is also presenting an opportunity, as will be argued in Chapter 5.

3

The world-economy and America's regions: from competition to dominance to volatility

The history of American involvement with the world-economy is also a history of regional growth and decline. The United States did not begin with a single core area. The concept of a federation of equal states also worked against the quick emergence of any one as central and dominant. Moreover, the opening of the western lands to settlement by Europeans led to a dispersal of the population and the creation of new regions. Thus from the start there was a notable lack of economic and political integration in the United States. Regions on the Atlantic seaboard with distinctive economies and cultures competed for dominance over the lands and wealth of the interior. This *competition*, resolved by the Civil War and the industrialization of the Northeast, created a long period in American history of political–economic *domination* by the businesses and politicians of the Northeast. This has now come to an end. Since the 1940s, but especially since the late 1960s, the western and southern regions have experienced much higher rates of economic growth and increased political influence relative to the Northeast. As this pattern is still emerging it is hard to say what the final outcome will be. Hence the characterization of this period as one of *volatility*.

One of the legacies of the past twenty years is a heightened awareness of apparent conflicts between different regions of the country. The immediate cause appears to be well-advertised shifts in population, economic activity and political influence between regions that have picked up pace since the late 1960s. These shifts and their connections with the evolving world-economy are the subject of Chapter 4. The concern here is to place these recent interregional conflicts in historical perspective.

To an extent that is more true of America than most countries, its history can be understood only as a composite of the history of separate regions. The continental magnitude of the country and the conditions under which it was settled have made this inevitable. A large part of American history, as Chapter 2 indicated, *is* the history of the progressive geographic expansion of people into diverse climatic, topographic and natural-resource environments.

The significance of America's regions

It is surely no accident that the most influential single force in American historiography has been Turner's hypothesis that one should look to life on the frontier for a deeper understanding of the forces shaping American institutions and Americans themselves (Turner 1920). Turner's frontier thesis, first proposed in 1893, shaped the research interests and strategies of several generations of American historians. While some have concentrated on the distinctive features of different frontiers – the Great Plains, the Rocky Mountains, the Pacific Northwest, for example – others have focused on the series of conflicts and compromises involved in a federal system comprising divergent and often centrifugal regional-interest groups.

The westward movement was critically important to the federal structure because the organization of 'territories' into 'states' constantly threatened to upset the balance of power within that structure. From the Missouri Compromise of 1820 to southern secession in 1861 the central political issue in the United States was regional domination of the federal government.

Even on issues that did not appear to be explicitly regional in substance, well-defined regional positions developed that reflected the interests of the *dominant* groups of each region. The essential cause then, as now, in the context of an essentially capitalist culture, was the balance of regional and local economic costs and benefits. If one considers, as will be done in more detail later, the three major, well-defined *antebellum* regions – a manufacturing Northeast, an agricultural West and a cotton-producing South – one could show that, on the four major economic issues of the tariff, internal improvement, alienation of public lands, and banking and credit policy, calculations of regional incidence dominated the pattern of congressional voting behavior. For example, the southern states raised the 'doctrine of nullification,' the view that states could reject and fail to enforce federal legislation of which they disapproved, many years before the Civil War and quite independent of the issue of slavery. One issue to which they applied it was the tariff legislation approved by Congress in 1824 and 1828. The South raised the doctrine of nullification because high tariffs, such as those approved, were a threat to the interests of southern planters. High tariffs might have benefited a Northeast anxious to encourage industrialization. To a South tied to European markets for its staple cotton exports and as a source of low-cost industrial imports, high tariffs were the 'abomination' southern spokesmen characterized them as at the time.

However, not only economic differences separate America's regions. Distinctive regional economies have been associated with distinctive regional cultures. As the economies have changed, the cultures have changed too, but more slowly. In addition, the impact of the world-economy upon America's regions through immigration waves and settlement preferences – *where* was booming at the time of arrival – cannot be neglected. The peopling of America by numerous ethnic groups and fundamental differences between 'founder' groups – compare, for example, the Puritans of New England with the Quakers of Pennsylvania – have created regional complexes that often defy neat correlation with contemporary economic interests (Elazar 1972).

Over time some of the issues that divide regions from one another have changed dramatically. If in the ninteenth century trade and tariffs were the central concerns, in the 1980s the allocation of federal government expenditures seems to excite the most interest. However, it is undeniable that some basic interregional differences became established in the nineteenth century that remain of importance today. For example, commercial agricultural specialization produced by a combination of such factors as rainfall, topography, soil chemistry, temperature variation and transportation accessibility led to well-defined patterns of regional specialization, so that it is possible to identify cotton belts, wheat belts, corn belts, tobacco belts, and so on. Farmers and congressmen in these regions still become excited about the same issues their grandfathers did eighty years ago – tariffs and bank loans, for example – as well as more recent ones like grain embargoes, price supports and new crop varieties. Others, especially the populations of regions in which these are not pressing questions, could, literally, not care less (Bensel 1984).

Though some regional differences may be as real today as they were a hundred years ago and new differences have arisen, there are certain respects in which regions have converged. One important respect is personal income levels. Economic historians have estimated personal income differentials by region as far back as 1840, when the average income in the Northeast was a third higher than the national average (see Table 3.1). This is no longer the case. Since 1880 regional per

Table 3.1. *Personal per capita income in each region as a percentage of the US average, 1840–1980*

	1840	1860	1880	1900	1920	1930	1940	1950	1960	1970	1977	1980
United States	100	100	100	100	100	100	100	100	100	100	100	100
Northeast	135	139	141	137	132	138	124	115	114	106	105	105
North Central[a]	68	68	98	103	100	101	103	106	101	100	100	100
South	76	72	51	51	62	55	65	72	80	85	90	86
West	—	—	190	163	122	115	125	114	105	101	103	103
Main deviation (%)				37			28			12		9

[a] North Central is equivalent to what was the 'West' before the Civil War.
Sources: Easterlin 1961: 528; ACIR 1980: 11; US Bureau of the Census 1981.

capita incomes have converged. Convergence was especially marked between 1930 and 1950, perhaps as a result of New Deal programs and the geographical spread of investment in defense industries. But interregional population movements – poor people moving from the South to the Northeast and the Far West and richer people moving in the opposite direction – and interregional capital flows – especially flows from the Northeast to other regions – have played important long-term roles. What is especially noticeable is that the Northeast dominated from 1860 until 1940 but has been in relative decline since then. This does not mean that the rest of the country has gained at the expense of the Northeast. In fact, excluding the period after the Civil War in the South, relative shifts in income levels have all taken place in the context of absolute increases in per capita incomes (ACIR 1980). What this means is that since 1940 other regions have had incomes growing at rates faster than the rate of income growth in the Northeast. However, even with convergence there are still important disparities between regions, particularly between the South and the others (see Figure 3.1).

Some commentators, noting trends towards regional convergence, have argued that 'regionalism' or 'sectionalism' was once a feature of

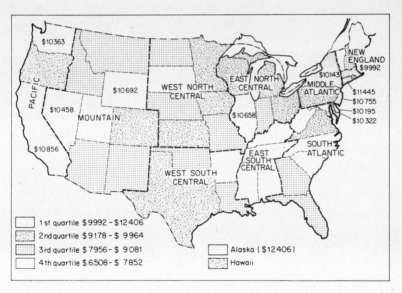

Figure 3.1: Per capita personal income by state and region, 1980 (US Bureau of the Census 1981)

the United States but has been displaced by a pervasive 'nationaliz-ation' (e.g. Boorstin 1965; and Lipset 1966). The history of the United States, they argue, is that of national patterns displacing regional and local ones. But this is to confuse a process with an outcome. It is undeniable, as Chapter 2 has attempted to demonstrate, that particu-lar places within the United States have become less and less isolated from one another and the world-economy. However, a world-economy has *always* provided the backdrop for regional definition and interaction. Regions never did 'define themselves,' so to speak, in isolation from wider processes of economic and political interaction. Moreover, the impact of global and national processes has always been to *create* regional distinctiveness rather than to displace it. For example, the economic–geographical 'fix' to crises of profitability has been to switch investments from some places to others. As Bluestone (1972: 66) remarks:

Those who control capital resources in the economy will tend over time to reinvest in those particular profit lines, machinery, geographical areas and workers which promise the highest return on dollar investment. Conversely, investment will tend to decline in segments of the economy where potential expected profit is relatively low. The outcome is continuous growth and rela-tive prosperity in the former sector and relative stagnation and impoverish-ment in the latter.

One aspect of the critique of 'regionalism' – the idea that there are significant differences in socio-economic organization and political outlook between America's regions – concerns the difficulty of defin-ing regions. Typical approaches to definition are vague. Regions are seen as loose geographic units larger than a state and smaller than the nation, with some objective characteristics of homogeneity. One can, however, identify regions that make sense in the context of American history. This is the way the problem of definition is tackled here. But the idea of fixed or static regions, the staple of orthodox American cul-tural geography (e.g. Zelinsky 1964) and the 'regional sociologists' (Jensen 1965), is best avoided (see Kirby 1984). America's regions have changed in their shape, if not in their importance, over time. For example, what was in the days before the Civil War 'the West' is now 'the North Central,' a region that since the turn of the century has become more and more 'like' the Northeast, *except* insofar as agricul-ture has remained a more important part of its economic base than in the Northeast 'proper.' Today some commentators pair it with the Northeast as the 'Snowbelt.' It is certainly part of America's 'industrial core.'

Regional competition

From the colonial period until the Civil War America did not have a single dominant region. Indeed the history of America during this time is essentially one of regional competition. But the competition changed its form from the colonial period, when it was between the three regions of New England, the Middle Colonies and the Coastal South, to later periods when Northeast versus South and, finally, Northeast (and West) versus South came to predominate. It also changed its content. In the early years competition was over the conduct of trade and the impact of British legislation. After Independence, tariffs and competition in the interior of North America for land and resources came to dominate.

The seaboard regions

Easily accessible coastal areas and river valleys provided the first sites for European settlement in North America. Before 1660 settlement was restricted to a zone from what is now Maine in the north to Norfolk, Virginia, in the south (see Figure 3.2). In 1700 an area without European settlement still separated the new city of Charleston from a newly settled area south of Norfolk. By 1760, however, most of the coastal plain had been settled and there was a movement of population into the valleys of the Appalachians.

Most of the settlers made their living at farming. For many this was subsistence; they provided only for themselves. But for others, and in increasing numbers, production was for the market, and the world market at that. This was especially the case with agriculture in the southern colonies. In terms of value of output southern agriculture was dominant throughout the colonial period and well into the nineteenth century. At the outset of settlement an attempt was made to establish crops from which an exportable surplus could be derived. First came tobacco. Later came two other staples, rice and indigo, and, from the 1790s, cotton. These crops, grown because the soils, climate and socio-economic organization of the South gave a pronounced advantage to the region, tied the region into the world-economy largely in terms of export dependence. Indeed, this is what defined the region as distinctive.

The land from the Potomac to the Hudson came to enjoy a comparative advantage in the commercial production of food. This is a measure of the extent to which production for the market was dominant in the American colonies even before Independence. The kind of agricultural unit that became established in the Middle Colonies later became typical of the great food belts of the West. Individual farms of relatively small acreage were operated by a family with a few indentured laborers. Slaveholding was rare, especially when compared to the South, where the size of plantations and the availability of year-around work made slavery efficient. The small but growing cities of New York, Philadelphia and Baltimore, and the Caribbean, were the major destinations for the wheat and other cereal crops in which this region specialized.

On the western edge of this region, however, subsistence agriculture and trapping for the fur trade were more typical. As this frontier advanced to the west larger and larger areas were opened up for commercial agriculture. This represented both the cutting edge of American expansion into the interior and a model for the frontier settlement later established on a continental scale.

The third seaboard region was New England. Compared to the other two, commercial agriculture was of secondary significance. Fishing, whaling, wood production and iron production were important. New England's comparative *disadvantage* in commercial agriculture during the colonial period, however, was of long-term advantage. Rather than being tied through the export of staples to foreign markets, New England was able to stake its growth upon the development of a truly American market as immigrants and territorial expansion added to America's domestic potential.

The most important characteristic of colonial commerce was the predominance of the South as the producer of the major staples. These staples (tobacco, rice and indigo) were what Englishmen under the mercantile system wanted from their colonies. The trade of Virginia, Maryland, the Carolinas and Georgia was essentially a direct trade with Britain. There was some direct trade with southern Europe and the British Caribbean, but these were minor exceptions. The main path of colonial export commerce ran straight from southern ports to Britain.

The trade of the Middle Colonies and New England, unlike that of the South, involved more than a simple exchange of staple commodities for finished goods. Northern commerce is epitomized in the famous colonial 'trade triangles.' Best known of these was one that began as a two-way exchange of fish, timber, livestock and provisions, shipped from the ports of New England, New York and Pennsylvania, for rum, molasses and sugar in the West Indies. Molasses was converted into rum by American distilleries and this was used to buy slaves on the West African coast. Slaves were in turn brought to ports

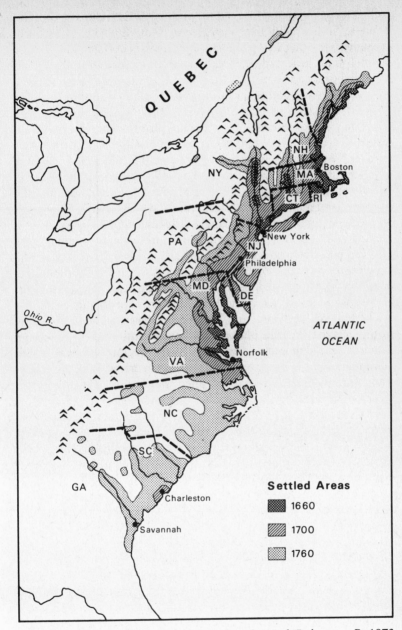

Figure 3.2: Seaboard settlement in the colonial period (Robertson, R. 1973:
52). Reproduced by permission of Harcourt Brace Jovanovich

such as Charleston and Richmond, or to the West Indies. In a second 'triangle,' a ship might take a cargo from Philadelphia or New York, exchange it in Jamaica or Barbados for molasses or sugar, and then go on to Britain and trade for textiles and iron manufactures before returning to its home port. A third 'triangle' involved trade with the Azores, Spain and Portugal as substitutes for the Caribbean.

This trade led to an emerging business class who, at least initially, relied on credit extended to them by their British counterparts. These resident merchants exerted influence out of all proportion to their number. There were probably no more than three hundred of them in 1770 residing in the port cities of Boston, Newport, New York, Philadelphia and Baltimore (Bailyn 1955). But they dominated the economic and political life of these cities and their hinterlands. Standing at the center of the trade triangles, the resident merchants came to dominate American resource allocation and capital accumulation. This capital would provide the future basis for American manufacturing industry. Perhaps most important, the political ambitions of these men, whose politics fed and supported their economics, assured a break with Britain when the break should prove profitable.

The trade of Maryland and Virginia, however, was dominated by British merchants, who placed their own agents in these American colonies. The dominant social class in this region was that of the plantation owners. Their basic interests lay in maintaining the export trade with Britain and covering their short-term debts to British merchants when prices were low by collateral in land.

The events of 1763 through to 1776 temporarily unified the interests and sentiments of the different regional elites and their regional populations. The resident merchants were directly threatened by much of the restrictive legislation imposed by the British government. The Sugar Act of 1763, for example, imposed a strict collection of a tax on the import into the American colonies of sugar and molasses from the British Caribbean. This was designed to protect British West Indian planters from the competition of New England rum-makers. But it disrupted the triangular trade and made resident merchants well aware of the constraints under which British merchants and *their* government would have them operate.

How, though, can the willingness of southern planters and small commercial farmers to join forces with the Northeast's merchant class be explained? There was again an important economic motivation. Just as the British government after 1763 inhibited trade, so British land policy placed a block in the way of agricultural expansion. Before 1763 British policy had encouraged the extension of agriculture into the interior. Rapid settlement was viewed as both stimulating trade and providing a block to French and Spanish territorial ambitions. By 1763, however, the need to build up the frontier against foreign powers had disappeared. Moreover, the fur trade was now under complete British control and required cooperation with the native Indian population rather than the liquidation increasingly favored by the colonists. A new land policy was therefore instituted to restrict land ownership and settlement by Americans to the west of the Appalachian Mountains.

The withdrawal of the 'free western lands' exerted considerable pressure on the planters in the tobacco colonies and on the frontier farmers in the Middle Colonies. In years of low prices many tobacco planters could meet their debts only through the sale of their western holdings, land they had obtained for practically nothing. British land policy tended to push the sympathies of this important group to the side of the merchant class of the Northeast.

Similarly, frontier farmers usually took an increasingly anti-British stand because of the new land policy. Many practiced a 'robber baron' style of agriculture requiring frequent moves into virgin territory as the soil was depleted. Others, as they switched to commercial agriculture, found themselves dependent upon an export of surpluses that was increasingly constrained by British trade regulations.

So for a number of years the three distinctive 'regional worlds' came together in common cause. However, the Revolution never did have the support of a substantial popular majority. Perhaps a third of the colonists remained loyal to the British connection. Many of them later moved to Canada or back to Britain. Another third did little or nothing to help the cause. Consequently only about one-third of the American population were 'Revolutionaries.' But they were present in all the colonies. The Revolution was a 'national' event.

The three distinct seaboard economies had associated with them equally distinct sets of cultural attributes. By 'cultural attributes' is meant 'systems of meaning' or complexes of values and beliefs that both inspire and justify practices and activities. Elazar (1972) labels the southern variety 'Traditionalistic,' the Middle Colonies' version 'Individualistic,' and the New England type 'Moralistic.' Elazar sees these as having particularly important political consequences. To him they are distinctive 'political cultures.'

The traditionalistic culture reflects a precommercial attitude that accepted both a hierarchical social order and personal relationships between the powerful and the powerless rather than impersonal or bureaucratic ones. The individualistic culture emphasizes the market place as the image of the 'good society.' It thus follows the Lockean emphasis on the centrality of private concerns and the need to limit government lest it interfere with the workings of the market. This is pre-eminently the ideology of the expanding world market and the one that came to predominate nationally in later periods of American history. The moralistic culture sees commitment to community and public life as morally superior to the decisions of the market place. By virtue of this orientation the moralistic culture produces a greater commitment than the others to active government intervention in economic and social life. But it is not collectivistic. At root it rests on making moral appeals to individual consciences. It would be incorrect to see this as a form of socialism, American style.

These regional cultures reflected both the ideas and practices brought by founder groups and beliefs that proved functional in the context of settling a 'new land.' The traditionalistic culture was the product of both transplanted 'establishment' attitudes from rural England and the realities of a slave-based plantation economy. The individualistic culture emerged from the origins of the Middle Colonies as a region of competitive commercial agriculture. The New England moralistic culture was a product of the Puritan ideals of its first settlers and their communal pattern of settlement.

Each of these regional cultures was in a fundamental sense incompatible with the others. But each was also altered later as it came into contact with the others and as the individualistic culture of the national commercial elite became predominant everywhere. The paradox of mainstream modern American culture, that it can be moralistic, deferential and individualistic at the same time, can be traced to such syncretism. However, it is clear that, at least in the 1760s, as many colonists were becoming conscious of themselves as 'Americans,' they had quite different senses of what this meant. Though in the aftermath of the War of Independence the alliance of northeastern merchants and southern planters survived to install an effective central government, it could not long continue once its original rationale was gone (Key 1964).

Northeast (and West) versus South

After the commercial interests of both the Northeast and the South had joined together to win Independence in 1776, and to create a Federal Constitution in 1787, they subsequently divided over how the new Republic would use its powers. To use Lynd's (1970: 60) words, they 'drifted, almost immediately into sectional cold war.' For example, in the new Congress that met in 1789, voting was along strict sectional lines. This was because 'The philosophy of the Anti-federalist, North and South, in 1787 had special charms for Southerners in 1790 because the issue of Federal interference with slavery had already appeared' (Lynd 1970: 56). Out of this conflict came the coalitions that provided the rudiments of the first American political parties.

After Independence the new United States was linked into the world-economy in two different ways. New England and the Middle Atlantic regions were dominated by commercial and trading interests, with a nascent manufacturing interest in New England. The South maintained its staple export base.

With the invention of the cotton gin the fibers of short-staple cotton could be easily separated from the seeds. A much larger area was then opened up for cotton production. Previously, cotton production had been limited to a narrow coastal strip in Georgia and South Carolina where an easily processed long-staple variety could be grown. The short-staple variety was much hardier and could be grown in a much larger geographical area. This both gave the South a new export crop much in demand in Europe and made the South even more important within the United States as a powerful exporting region.

Cotton-growing enterprises, based on the use of slave labor, were highly organized, labor-intensive units of production. As the plantations moved westward cotton became the dominant crop. This involved a growth in scale and an increase in the number of slaves. Most cotton production came to be carried out on estates with 30 slaves or more. In the alluvial lands of the Mississippi flood plain 200 slaves or more became the norm. High intensity of labor was achieved through the assignment of labor gangs to successive specific tasks within the cycle of cultivation, a process reminiscent of industrial production. Force was crucial to the system (Fogel and Engerman 1974).

Cotton-growing with slave labor was a profitable business. On average, slave owners earned about 10 percent on the market price of their slaves, which compares favorably with the returns on investment for New England textile firms in the mid-nineteenth century (Fogel and Engerman 1974). But growing cotton was not the cause of slavery in the United States, even though it was an important factor in its continuance into the nineteenth century. Slavery was an integral part of southern socio-economic organization. It was a social reality. However, demand for cotton, primarily from England, gave material strength to the system and facilitated its geographical expansion.

From a total output of 3,000 bales in 1790, cotton production rose to 178,000 bales in 1810, 732,000 in 1830, and 4,500,000 in 1860 (Fogel and Engerman 1974: 44). This growth of production was accomplished through a geographical spread of cotton-growing from the Atlantic seaboard into the Deep South along a climatic belt in which there was a minimum of 200 frostless days and sufficient rainfall for a successful cotton crop (see Figure 3.3). The spread of cotton-growing was paced by the steady advance of transportation by steamboat and, later, railroad, which brought even more distant areas into contact with the major ports. This expansion involved a major relocation of people. Between 1790 and 1860, 835,000 slaves were moved – mostly from Maryland, Virginia and the Carolinas towards Alabama, Mississippi, Louisiana and Texas. More than two-thirds of these slaves moved between 1830 and 1860 in one of the great forced migrations of all time. In the early nineteenth century, therefore, the slave-based economy of the South was not a stagnant, residual or dying one; it was dynamic and aggressively expansive.

The American slave population was largely self-reproducing in contrast to the slave population of the Caribbean. 'Of all the slave societies in the New World,' writes Genovese (1972: 5), 'that of the

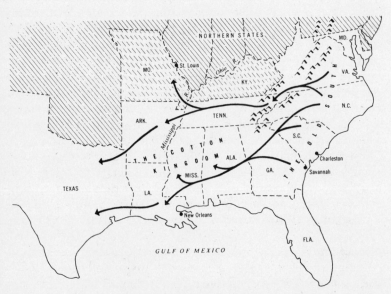

Figure 3.3: The spread of cotton growing, 1812–60 (Robertson, R. 1973: 116). Reproduced by permission of Harcourt Brace Jovanovich

Old South alone maintained a slave force that maintained itself.' The reasons for this are not clear. Slaves do not seem to have been systematically bred for sale (Fogel and Engerman 1974). Perhaps a lower incidence of disease, especially venereal disease, and higher levels of family formation and kin networks account for the difference (Gutman 1976).

Just before the Civil War cotton was indeed 'King' in the South – and more generally. As North (1961) has noted, it is difficult to exaggerate the role of cotton in American economic growth between 1800 and 1850. This great staple accounted for half the dollar value of US exports, a value ten times as great as its nearest competitor, the wheat and wheat flour of the Middle Atlantic. In the United States, cotton planters provided the raw materials for textile manufacturers in New England, who by 1860 were engaged in a major switch from wool to cotton. Small wonder that the aristocratic southern planter class could scarcely envisage a Northeast, or even a world, without their chief product.

Most white southerners were not slaveholders and the plantation was not the typical unit either; small, family-operated farms far exceeded plantations in number. This has led some commentators to downplay the significance of slavery in the South before the Civil War (Owsley 1949). But it was the slaves on the plantations who produced the bulk of the region's marketable surplus and thus paid for the South's imports from Europe. It was the labor of slaves that provided the resources to maintain mansions and Mississippi riverboats, cotton brokers and lawyers, and the way of life of the southern planter. In short is was slavery that maintained the culture of the region. Moreover, the majority of planters, agrarian and status-minded as they liked to portray themselves, were adventurously and ruthlessly bent on profit and reinvestment. That is, they were agricultural entrepreneurs in a capitalist society; their importance as a class resided in their ability to accumulate surplus for investment.

But there was always more to the South's defense of slavery than economics; there was also race. All the profits of the plantation cannot explain the tenacity with which the 'little people' of the South – the Europeans without slaves – came to the defense of the slave system, both before the Civil War and in the armies of the Confederacy. Their stake in slavery is probably found in the threat 'free' blacks would pose to their own marginal status and the psychological–sexual fears American blaks arouse in many white Americans to this day. Slavery was never simply a means of extracting more surplus; it was also a means of social control. Even non-slaveholders who opposed slavery for moral and economic reasons were often silenced by slavery's ability to control blacks (Olmsted 1861).

As the slave system expanded to the West, so too did the Middle Atlantic system of agrarian capitalism. Early settlements in Ohio were made by people who came down the Ohio River from Pittsburgh. They were joined by pioneers who, having stopped for a while in Kentucky or Tennessee, moved north and west into Ohio, southern Indiana and southern Illinois. But much larger flows in the early 1800s came directly from New England and the Middle states. They settled western New York, northern Ohio and Indiana, and southern Michigan. On the eve of the Civil War settlers were pushing into central Minnesota and and eastern Kansas (see Figure 3.4).

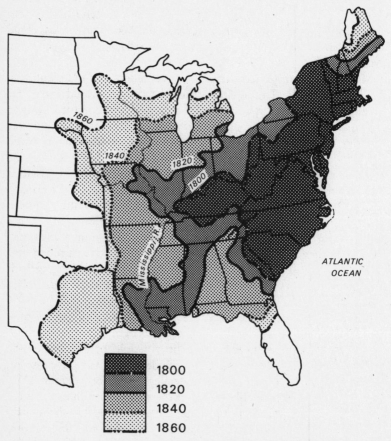

Figure 3.4: The moving American frontier, 1800–60 (Robertson, R. 1973: 114). Reproduced by permission of Harcourt Brace Jovanovich

Southerners, then, moving across the Ohio River, were the chief influence in the lower part of what was then called the Northwest. New Englanders were dominant in the Great Lakes region, especially once the Erie Canal allowed for easier movement, but they were joined by a major stream that originated in the Middle Atlantic states. For the most part, families moved singly, though sometimes large groups of fifty to a hundred would move together.

Throughout this period there was an ever increasing influx of land-hungry people from abroad. From 500,000 people in the 1830s this flow increased to 1.5 million in the 1840s, and to 2.5 million in the 1850s. German immigrants in particular tended to go directly to the West.

Whatever their place of origin, settlers were faced with a difficult and isolated life. The arduous work of clearing the land – it might take a generation to clear a farm of 160 acres – was compounded by social isolation. The proportion of the family's work leading to immediate consumption was limited by the necessity of providing for the future. Clearing, building and fencing were thought of as 'investments.' Early on it became part of everyday life to keep an eye on the market for homesteads and to have in mind a price at which to sell and move on (North 1961).

Even settlement on the prairies was not without its difficulties. Plowing heavy prairie sod was not easy. After 1845, with large increases in the world prices of grains and with new railroads running to the east, however, prairie farming became attractive. Indeed, the West, like the South, had its staples – wheat, corn and livestock – at a very early stage of settlement. Very early, also, meat-packing centers arose in the West focused on pork-processing. Pig-raising implied corn-growing. Corn could be grown very easily given certain minimal climatic conditions. But wheat was to be the greatest commercial crop from the West. As in the case of the South and cotton, the major determinant of the pace of westward expansion was the profitability of the major staples, in this case wheat and corn (Billington 1960).

The shift of wheat production was steadily westward. In 1850 New York and Pennsylvania were still the major producers but by 1860 Illinois, Indiana and Wisconsin had taken over. The major wheat-growing areas had yet to be finally established; but the movement of cash grain production was entirely away from the Northeast and to the West (North 1961).

The westward expansion seemed at the time, and to many people since, as the 'American dream' come true. There appeared to be land for the taking, and agrarian democrats in the Jeffersonian tradition looked forward to a nation of sturdy yeomen, servants to no man through their possession of land. Black slaves excepted, that is. But, of course, America was not 'a land without people for a people without land.' Land was occupied and used by native Americans. To make yeomen, these native had first to be dispossessed. To most frontier Americans there was little doubt that this was justifiable. Reaching back to colonial times, the fiction that Indian tribes were 'nations' fully capable of negotiating treaties to cede away their 'occupancy titles' was a suitable mask behind which forced expropriation could take place without moral anguish. As long as space remained to which defeated and demoralized tribes could be removed the 'treaty system' worked effectively. Under an elaborate system of treaties, supplemented by the Indian Removal Act of 1830, defeated eastern tribes and tribal remnants were transplanted to reservations set aside for them in the West.

Between 1820 and 1840 three-quarters of the 125,000 Indians living east of the Mississippi were removed; between one-quarter and one-third of all southern Indians died. By 1844 less than 30,000 Indians remained in the eastern United States, most of them located around Lake Superior. The entire process of forced migration was accompanied by paeans of praise extolling the victory of civilization over savagery (Young 1958). The successful conclusion of the operation realized President Jackson's hope that Indian land could be speedily brought 'into market' (quoted in Rogin 1975: 174).

As the West was settled it emerged as a section to be reckoned with in national political life. By the late 1820s, although some differences continued to exist between the commercial interests of the Middle Atlantic and the new manufacturing interests of New England, there were three sections still. But now they were the South, the Northeast and the Northwest (or West). The situation at that time has been summarized rather dramatically by Billington (1960: 29): 'The settlement of the trans-Appalachian frontier brought the United States face to face with a terrifying problem: how could an industrial Northeast, a cotton-growing South and a small farming West live side by side in peace?'

Ultimately, of course, they could not. Until the railroad irrevocably tied the West to the Northeast there was no resolution. However, for forty years the national political arena in Washington was dominated by the competing claims of Northeast and South in the West. Regional rivalries centered upon four areas of federal responsibility: public lands, trade tariffs, banking policy and transport improvements (Billington 1960; Schlesinger 1945).

From the ratification of the Constitution, when the Atlantic states gave up their claims to western territories, the federal government was faced with the problem of disposing of public lands. This included conditions of sale which related in turn to the type of farming and inevitably brought up the issue of slavery. Hence, land policy provided the basis upon which conflict over the extension of slavery into the West rested. In the Missouri Compromise of 1820 a tentative resolution was reached through the balanced admission of 'free' and 'slave' states. In this conflict the West sided with the Northeast, although they were divided over the question of land sales as a source of federal revenues.

On trade tariffs regional rivalries were more complex. Billington (1960) shows how Protection Bills from 1824 until 1846 elicited sectional responses, with the South favoring free trade to maintain its role as a cotton provider to Britain. On this issue the Northeast was divided. New York's commercial interests supported low tariffs while New England's manufacturing interests favored protection. Attitudes in the West shifted over time from one of alliance with the South to one of growing support for protection.

The third area of federal policy in which the regions clashed was banking policy. In the 1830s there was a cleavage between the Northeast on the one hand, in which there was support for the national central bank, the second bank of the United States, and elsewhere on the other hand, where the bank was seen as an instrument of the northeastern 'establishment.' This was the beginning of a major theme in nineteenth- and early twentieth-century American politics: Populist opposition to the Northeast's dominant financial position in national economic life.

The fourth and final area of federal policy over which the sections were in competition was transport improvements. Revenues from higher tariffs and the sale of public lands provided a fiscal surplus. One use for surplus funds was to promote transport improvements. Some funds went into road improvements such as the famous National Road linking Cumberland, Maryland and Wheeling, Virginia, thus linking the Atlantic seaboard to the Ohio Valley. It is in the realm of water transport, however, that competition between the Northeast and the South was mostly seriously engaged, as Billington (1960: 329) puts it, 'to tap the surplus-producing areas developing in the trans-Appalachian country as the self-sufficient economy of the frontier gave way to crop specialization during the 1820s.' Initially the South was in the most advantageous position as the steamboat enabled the region to exploit its natural advantage: the Ohio–

Mississippi river system. The Northeast's response was to create a system of canals linking the Hudson River to the Great Lakes. After the construction of the Erie Canal the balance was restored to the competition.

The transport competition between the regions was conclusively settled with the advent of the railroad, as argued in Chapter 2. The increasing complementarity between the economies of the Northeast and the West made railroad investments between these regions safer and better. Billington (1960) has shown that the West as a surplus region and the Northeast as a deficit region formed natural partners to the detriment of the South. This situation was reinforced by foreign-trade patterns that emerged after the repeal of the British Corn Laws in 1846. Before then most of the American grain surplus went to the West Indies and Latin America via New Orleans. After 1846 'the center of the export trade shifted to the Atlantic ports closest to the British consumer' (Billington 1960: 400). The final outcome was the emergence of a Northeast–West economic axis (see Figure 3.5). Across a range of commodities an integrated *northern* economy was emerging that provided the beginnings, after the Civil War, for the American manufacturing belt. By 1860 this area was discernible as the economic core of the United States. Sectional conflict was thus reduced to North versus South, core versus periphery. The scene was set for the Civil War.

In the national political arena alliances reflected these economic changes. The positions of the three sections (West, South and Northeast) can be summarized as follows (Billington 1960: 353):

	Public land	*Tariffs*	*Banks*	*Transport*
West	low price	protective	hostile	federal support
South	high price	low tariff	hostile	no support
Northeast	high price	protective	friendly	federal support

With hindsight the weak position of the South is obvious. On only one issue, federal land policy, did the Northeast and the South coincide. For the West this was the most important issue and after 1845 it was this that brought the Northeast and the West together to oppose the extension of slavery, upon which the growth of the South depended.

Lynd (1970) provides a succinct summary of what happened in terms of major themes in American historiography. The War of Independence and the Constitution brought together sectional interests in a common cause to create a new state. Once this was achieved, regional rivalries, although managed cleverly for many years, became endemic as different regional interests competed to decide what sort of society America would become: 'A showdown was postponed . . . because each sectional society expected to augment its power from new states to be formed in the West. What Turner's frontier thesis explains is why Beard's second American Revolution (the Civil War) was so late in coming' (Lynd 1970: 60).

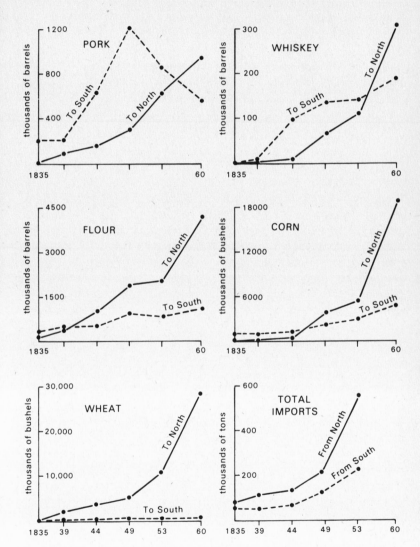

Figure 3.5: Trading links of the 'old' West (Northwest) (Archer and Taylor 1981: 75). Reproduced by permission of Research Studies Press (John Wiley and Sons)

Regional dominance

The entire balance of regional interests was fundamentally altered by the Civil War. The North was now firmly in control over the American economy and polity. Down until the 1890s, however, a new West (beyond the Mississippi) was still in the process of settlement, a new urban–industrial order was being forged, and the United States was still largely prey rather than predator within the world-economy. Only after the depression of the 1890s was northeastern dominance established as a fact, a fact that lasted until World War II.

The rise of an American core

In the period from the Civil War until the 1890s sectional competition involving the South was replaced by the growth of the North in a process sometimes termed the 'nationalization' of the American economy. Transport innovations, railroads in particular, economically integrated the 'old Northwest' or West with the Northeast to create a single 'North' by the 1850s. This integration culminated in the North's support for Lincoln in the presidential election of 1860 that precipitated the Civil War.

However, as this old West was being absorbed into a new regional alliance, a new West was developing beyond the Mississippi. In 1850 California had been admitted to the United States and plans for a transcontinental railroad were not slow in following. But the politics of sectional stalemate prevented much progress on such a project. The Civil War ended the deadlock. In 1869 the line was finished. This signalled the start of a nationally integrated economy. By 1883 there were four transcontinental railroad links. The northern routes were primary. This signalled that equally surely the new national economy revolved around the economic leadership of the North.

In the second half of the nineteenth century the United States was transformed into a major industrial state and the process of industrialization was concentrated in the North. Urban growth associated with the new industrial growth produced a manufacturing belt stretching from the Atlantic to Lake Michigan. This area was not one continuous industrial complex, but was characterized by a number of very large urban centers with a level of manufacturing industry much higher than the national average. In 1870, 77 percent of all American manufacturing employment was located in the belt even though only 56 percent of the nation's population lived there. By 1910 this concentration

Table 3.2. *Proportions of the total labor force employed in manufacturing, by region, 1870–1910*

	% of total		
	1870	1890	1910
New England	44.1	47.9	49.1
Middle Atlantic	32.1	35.8	39.8
Great Lakes	19.7	25.0	33.2
Plains	14.9	16.4	20.0
Mountains	12.9	20.0	20.5
Far West	18.2	22.6	27.0
Southeast	7.5	10.5	14.5
Southwest	7.0	9.1	12.4
United States	21.1	24.5	27.9

Source: Perloff *et al.* 1960: 172–83.

had declined somewhat (68 percent to 48 percent), but it remained at that level until 1940 (Berry and Horton 1970).

In 1870 over one-half of the country's population lived in New England and the Middle Atlantic states. Between 1870 and 1910, although these areas maintained a vigorous rate of population growth, there was an even more rapid growth of population and manufacturing employment in the western part of the Middle Atlantic states and in the vicinity of the Great Lakes (see Table 3.2).

By 1910 a now clearly defined manufacturing belt (see Figure 3.6) supported 34 of the 50 cities with more than 100,000 inhabitants and 14 of the 19 cities with more than 250,000 inhabitants. Cities such as Cleveland, Pittsburgh and Detroit had grown especially quickly. In 1870 not one of these cities had a population greater than 100,000, but by 1910 Cleveland and Pittsburgh each had more than 450,000 inhabitants (Ward 1971: 41).

The highly integrated urban system of the manufacturing belt was surrounded by a system of peripheral centers in resource regions that served the manufacturing core. New Orleans, San Francisco and Minneapolis became especially important commercial 'gateways' to the Mississippi Basin, the Central Valley of California, and the northern Great Plains, respectively. These were supplemented later by further gateway cities such as Los Angeles (southern California), Kansas City (the central Great Plains), Seattle (the Pacific Northwest), Dallas and Houston (Texas) and Phoenix (the Southwest). In each case, as Berry and Horton (1970: 35) put it:

the basic conditions of regional growth were set by the heartland. It served as the lever for successive development of newer peripheral regions by reaching out to them as its input requirements expanded, thus fostering economic specialization of regional roles in the national economy . . . Flows of raw materials inward, and of finished products outward, articulated the whole.

This core–periphery structure actually contained two distinctive peripheries. The North constituted a core, but the periphery was not united. Rather, it consisted of a West into which vast investments were placed, including payments in the form of pensions to Civil War veterans (Union only), and a South which was drained of its resources rather than developed through investment. After the Civil War the South sank into a tributary condition as the most backward section of the national economy. Despite the tremendous westward shift of the population, the South's share of primary-sector employment (agriculture, mining, etc.) still increased (Perloff *et al.* 1960). The result was a dynamic West and a backward South. The South had lost a war and paid the penalty by becoming dependent on the North and by missing out on the western bonanza.

By the 1890s a distinctive three-region core–periphery geographical structure characterized the US economy. This structure could be exploited by the dominant business groups in the North to maintain their national hegemony (Archer and Taylor 1981). Thus when Populist groups in the South and West arose to challenge the protectionist policies favored by northern business, the western groups were easily mollified by a new national policy of overseas expansion in which the dynamic West would share. This produced a dramatic reversal between the presidential elections of 1896 and 1900. In 1896 Populist groups almost managed to have their candidate, Bryan, elected, but by 1900 McKinley, the victor in 1896, had managed to disengage the western elements from the southern ones (Glad 1964; Williams 1969). Hollingsworth (1963: 130) links the regional politics of the 1890s with the 'solution' of overseas expansion:

It was no accident that a jingoistic spirit became particularly intense in the same regions in which the demands for agrarian reforms had recently been so pronounced and in which a discontented constituency, frustrated by Bryan's defeat, was vociferous. Many agrarians, discouraged in their struggle for domestic reforms, embraced the opportunity to carry their crusading ardor to the Cuban battlefield, when the conflict was viewed as a struggle between tyranny and popular rule.

The birth of the Farmers' Alliance in the late 1880s, in a period of falling crop prices and rising rents, had signalled a radicalization of agrarian protest in the United States. Especially in the South, where the *ancien régime* had been recast into the debt servitude of the share-cropping system, the Alliance, by its unprecedented feat of uniting black and white tenants had become a subversive force of revolutionary potential (Goodwyn 1978). By 1894 a more conservative and anti-labor bloc of wealthier farmers from the Great Plains was displacing the leadership of the radical southern and southwestern Alliance men (Montgomery 1978). Rather than focusing on tenant issues, this group preferred to campaign for silver rather than gold as an international currency standard. The silver campaign amounted to an economic nationalism whereby the 'American goldbugs,' were seen 'as the willing accomplices of British control of the world marketplace' (Williams 1969: 306). Silver was preferred 'because it would place the farmers of the South and Northwest upon equal footing in the market of the world' (Williams 1969: 364).

The 'Populist moment' passed, then, without long-term effects.

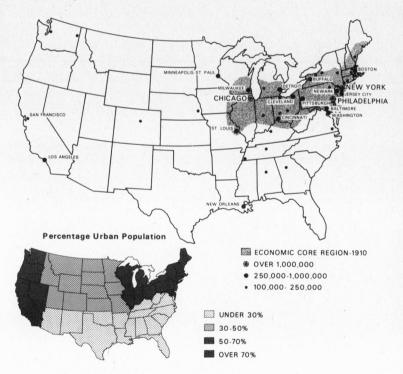

Figure 3.6: Urbanization in the United States, 1910 (Ward 1971: 42–3). Reproduced by permission of Oxford University Press

Inside the core the 'cultural style' of Populism was of limited appeal. Its strong affinities for prohibitionism and state education reproduced classic nativist and anti-immigrant motifs. The core was the region *par excellence* into which foreign immigrants poured. So at the same time that European workers were becoming politically more engaged than ever, the American working class was undergoing an electoral demobilization. In particular, ethno-religious alignments within the core, individual mobility as a solution to oppressive working conditions, and divisions within the agrarian Populism of the periphery, prevented the emergence in the United States of a social democratic or labor party along European lines (Davis 1980). Rather, the Republican Party, the party of northern business, dominated in the core, and the Democratic Party represented the outsiders of the periphery (Bensel 1984). Only with the coming of the New Deal in the 1930s did participation in elections by workers increase and was Republican dominance of the core ended.

By the turn of the century and after the scare of 1896 the northern core was overwhelmingly dominant within the national economy. But economic dominance rested on more than electoral votes. National income during the period from the Civil War until the 1890s was systematically redistributed from the periphery to the core. In particular, the South helped pay for northern industrialization either through higher prices for domestic goods or through duties on imported products. The export of southern cotton underwrote northern economic expansion rather like collective agriculture was later to underwrite Soviet industrial expansion.

Economic subordination of the South, however, was a prerequisite not only for northern political domination, but also for American overseas expansion. Unlike the industrial core or the agricultural West, the South and mountain West produced raw materials that required *industrial* markets either in the United States or abroad. Colonies or undeveloped markets such as China were attractive to producers of foodstuffs and industrial goods but not to miners or southern cotton brokers. Instead of the imperialism favored by the northern core, these peripheral interests preferred free trade that would remove them from bondage to particular markets. This could only occur with the abolition of tariff barriers and competition between *alternative* industrial markets (Williams 1969).

Industrial core versus resource peripheries

From 1900 until 1940 the United States established itself as a major actor in the world-economy. Or, more to the point, the northern industrial core did. In the 1890s the debate over overseas expansion was won by the core. America was emerging from a semi-periphery role within the world-economy and directly raising a challenge to British global hegemony.

Inside the United States the three-region core–periphery geography persisted and was consolidated. Perloff *et al.* (1960: 22) could assert that 'By 1910 the dominant regional pattern with which we are today familiar had already taken shape.' In 1957, for example, the manufacturing belt contained nearly the same proportion (46 percent) of the population as it had in 1900 (Perloff *et al.* 1960: 22). After 1940, however, a transformation of the internal geography of the United States was under way. Perloff *et al.* (1960: 22) cannot be faulted. They have an escape clause: 'the apparent fixity of the twentieth-century regional structure can be overstated.'

This period is largely one of returns to cumulative advantage – or disadvantage. Industry continued to be concentrated in the manufacturing belt. Industrial agglomeration provided its own reward. It lowered the costs of production for many firms by making readily available a wide variety of materials, labor and services. But over time locational patterns did change. In particular the Middle Atlantic region and New England became relatively less important. Generally, new manufacturing industries were located further west to take advantage of nearby material inputs (especially iron ore and coal deposits). This was especially true of the steel industry, which shifted towards lakeside sites in the Great Lakes region. However, agglomeration economies are not an absolute or transcendental advantage, although they are often portrayed as such. They must be defended politically. US involvement in World War I and the stimulation of mass consumption by the easing of credit restrictions in the 1920s helped. But it was political domination by the business interests of the core that reinforced the economic advantages of the manufacturing belt.

World War I was especially important. Core business interests were well served by the vast mobilization and coordination of industrial capacity during the war. The government suspended anti-trust legislation and encouraged corporate consolidation. Moreover, the large financial institutions of the Northeast profited directly from extending credit to combatants and financing war production. Not surprisingly, as Bensel (1984) has shown, Republican congressmen and some of their Democratic counterparts in the northern core supported an aggressive US foreign policy and military expansion. Elsewhere there was less enthusiasm. Indeed, some congressmen from the Great Plains and the old West held to a so-called 'isolationist' position in foreign affairs. This reflected both the remnants of the Populist movement and hostility to the British-style imperialist policies pursued by core Republicans – the eastern 'money power' (Billington 1945). Their view was shared in part by representatives from other peripheral regions (Bensel 1984).

From 1900 until 1933 economic changes and public policies pitted the interests of northeastern capital against the economic imperatives of periphery agriculture and mineral production. The periphery political party, the Democrats, only controlled Congress when it could successfully exploit usually latent class conflict in the core. Since the Democratic Party was dominated by periphery elites this alliance was usually short-lived.

It was also difficult for peripheral groups to make common cause *without* the complication of attracting support from the core. The agrarian periphery cultivated a wide variety of crops and raised livestock for domestic and foreign markets (see Figure 3.7). Three crops in particular stood out in terms of general economic importance: cotton, wheat, and corn (especially important for its role in hog production). Throughout the 1920s regional specialization in these crops increased (Soule 1968). The economic interests of rural hinterlands in these regions also controlled the politics of local urban areas because

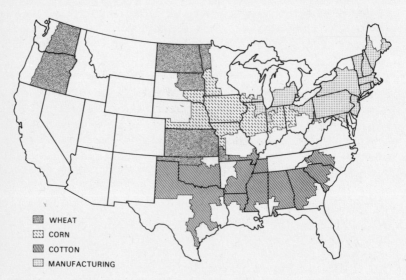

WHEAT
CORN
COTTON
MANUFACTURING

Figure 3.7: The major specialized economic territories of the United States, 1925 (Bensel 1984: 138). Reproduced by permission of University of Wisconsin Press

of their role as resource gateways. Each of the crops had a distinct relationship to the world-economy. Of particular importance, cotton was still a major export crop. Its production did not involve the tremendous overproduction and periodic price busts of the cereal grains. Most fundamental of all, however, by the 1920s the American economy was organized entirely around the interests of the northern core. There were increasingly unfavorable terms of trade between commercial agriculture and the American industrial economy (Heady 1962). H. R. Smith (1955: 547) saw the problem as follows:

The natural course of events in an industrializing economy is for exports to become increasingly manufactured products while imports are becoming increasingly of the raw material variety. This follows from the fact that industrialization implies a progressive cheapening of the factors of production most essential to factory production, and a progressive increase in the cost of those factors most needed in raw material producing industries. Artificially to stimulate agricultural exports would unquestionably be tantamount to slowing down this evolution.

It became a question of every crop for itself.

The New Deal revolutionized the American political economy. In so doing it made possible a pluralistic political system in which the working class of the core, represented by organized labor, could permanently participate in the inner councils of the periphery (Democratic) party. As Hawley (1966: 197) notes in his classic study of the New Deal, this was a product of government intervention in the economy.

What had been an essentially monolithic economy in 1929, one dominated for the most part by the business–financial element, was being converted into something basically different, an economy of great countervailing forces in which organized groups fought each other for their respective shares of the national income and appealed to political power to aid them in this struggle. For agriculture the government had moved into the marketplace and by the use of public power had given the farmers the advantage of corporate organization without forcing a collectivization of actual operations. For labor governmental intervention in the bargaining process had not been so complete, but labor organizations and the growth of labor power were certainly encouraged to the extent that the Administration [under Roosevelt] maintained a friendly attitude, established minimum standards, absorbed surplus labor, required recognition of unions, and restrained a number of employer practices that had been used to break unions in the past.

Federal intervention, therefore, gave commercial agriculture and organized labor a political base to counter the strength of high finance and industry. Ultimately this meant that the peripheral economies came to possess certain advantages they had lacked previously and the increased power of organized labor weakened business dominance in the core.

The collapse of the American economy thus allowed the periphery-based Democratic Party to expand into the industrial core. The result of this expansion was the New Deal coalition. The anomalous loss of influence by the core industrial and commercial elite created a national power vacuum that allowed this to happen. This did not last for long. The New Deal coalition was shaky from the start. Core and periphery factions discovered that their interests and goals were often incompatible. More importantly, the mobilization of the American economy during World War II checked the New Deal redistribution of power. However, out of the politics of the New Deal emerged a widely accepted predisposition in favor of central government intervention in the economy. The critical question became: intervention for whom?

Regional volatility

Before 1933, the Republican Party had continuously used regional conflict as a strategy for retaining national power. Until the late 1960s the Democratic Party sought to minimize regional competition to the same end. Its approach to retaining power was an interventionist central government that promised something to everyone by letting business 'loose.' The increased power of organized labor arising from the New Deal, however, and resource advantages (particularly energy-related ones) in the periphery were to make the industrial core, *especially* the heavily unionized industries, less and less attractive for business investment in the decades after World War II.

The 'aging' of the industrial core was also involved. Old industrial plant in industries with declining profit rates relative to newer industries presented a barrier to future growth. Many of the advantages the industrial core once had in terms of infrastructure and multiplier effects became disadvantages. In particular, high levels of unionization retarded the introduction of new technological processes, and the costs of maintaining aging public facilities – roads, sewers, railroads – were ones business was anxious to avoid.

From World War II until 1967 the decline of the industrial core was gradual. It was also a relative loss of economic activity rather than of economic power. Much of the new industry in the historic periphery consisted of branch plants or facilities of locally owned firms that had expanded after being taken over by major national firms (Jusenius and Ledebur 1977). Capital and decision-making remained concentrated

Table 3.3. *Yearly percentage of population growth by region,*
1941–80

	1941–50	1951–60	1961–70	1971–80
Northeast	1.09	1.56	1.01	−0.01
North Central	1.56	1.76	1.06	0.33
South	2.61	2.59	1.38	1.75
West	3.75	3.50	2.37	2.08
United States	2.14	2.26	1.36	1.02

Source: Leven 1981.

in northeastern cities (Pred 1974). In the 1940s and 1950s New York became even more important as a corporate center than previously (Goodwin 1965; Elazar 1968).

The process whereby industrial production spread beyond the confines of the industrial core began on a large scale during World War II. The War Production Board had as one of its objectives the dispersal of new capacity to 'uncongested areas.' This also stimulated population movement to the regions where the new industries were located (see Table 3.3). This movement was basically to the West but it had tilted by the 1960s towards the South (see Figure 3.8). The result was to enlarge permanently the periphery's share of national industrial capacity (see Table 3.4). This relocation policy was to have a built-in multiplier effect in the post-war period. Most of the new industrial capacity developed during the war was 'defense'-related. Thus, after the war and with the growth and maintenance of a significant war economy in the United States, it was these places that benefited from the vast government expenditures on military hardware.The key to the growth of southern California, for example, is not to be found in Hollywood but in Washington DC (more accurately, across the Potomac in the Pentagon) (Sampson 1977).

The 1950s and 1960s were decades of explosive growth in the US economy as a whole. Involved in this expansion were two important changes that had major geographical implications: the development of new transport and communication technologies and the continued centralization of capital that provided the means to exploit the new technologies (see Chapter 2). The construction of the inter-state highway system, the development of computers and long-distance communications systems, and the growth of air travel, provided a 'permissive' technological environment in which businesses could decentralize their manufacturing operations yet maintain central

Table 3.4. *Comparison of industrial production before World War II with war-time facilities' expansion*

	Pre-war private economy		War expansion	
	1939 value of production ($ million)	Percentage of total US production	Value of production added, 1940–5 ($ million)	Percentage of total added US production
New England	3,877	9.8	1,101	5.1
Middle Atlantic	11,788	29.8	3,941	18.2
East North Central	12,461	31.5	6,773	31.4
West North Central	2,176	5.5	1,688	7.8
South Atlantic	3,600	9.1	1,551	7.2
East South Central	1,345	3.4	1,248	5.8
West South Central	1,305	3.3	2,544	11.8
Mountain	435	1.1	818	3.8
Pacific	2,571	6.5	1,938	9.0
Total United States	39,558	100.0	21,602	100.0[a]

[a] 1940–5 total does not include 3,556 in productive capacity that was not distributed among the regional categories.
Source: Bensel 1984: 182.

Figure 3.8: The geographic center of US population, 1790–1980 (US Bureau of the Census 1981)

Table 3.5. *Percentage of sales accounted for by the four largest producers in selected manufacturing industries, 1947–72*

Industry code	Industry	Percentage of sales	
		1947	1972
2043	Cereal breakfast foods	79	90
2065	Confectionery products	17	32
2067	Chewing gum	70	87
2092	Malt beverages	21	52
2211	Weaving mills, cotton	18[a]	31
2254	Knit underwear mills	21	46
2279	Carpets and rugs	32[b]	78
2311	Men's and boys' suits and coats	9	19
2337	Women's and misses' suits and coats	3[a]	13
2421	Sawmills and planing mills	11[a]	18
2771	Greetings card publishing	39	70
2829	Synthetic rubber	53[a]	62
3211	Flat glass	90[a]	93
3312	Blast furnaces and steel mills	50	45
3421	Cutlery	41	55
3511	Turbines and turbine generators	90[b]	93
3555	Printing trades machinery	31	42
3585	Refrigeration and heating equipment	25	40
3624	Carbon and graphite products	87	80
3633	Household laundry equipment	40	83
3636	Sewing machines	77	84
3641	Electric lamps	92	90
3661	Telephone and telegraph apparatus	90	94[c]
3674	Semiconductors and related devices	46[b]	57
3679	Electronic components	13	36
3692	Primary batteries, dry and wet	76	92
3711	Motor vehicles and car bodies	92[d]	93
3724	Aircraft engines and engine parts	72	77
3743	Locomotives and parts	91	97[d]
3861	Photographic equipment and supplies	61	74
3996	Hard surface floor covering	80	91

[a] 1954.
[b] 1963.
[c] 1970.
[d] 1967.
Source: Bluestone and Harrison 1982: 120.

control. These technologies made it possible for businesses to take advantage of new sources of cheaper and tractable labor in peripheral locations both within the United States and beyond its borders (Bluestone and Harrison 1982).

The development and use of these new technologies were stimulated by the large firms that came to control significant sections of American industrial capacity in the aftermath of World War II (see Table 3.5).

Firms that can dictate price, free of any competitive pressure, and restrict output by barring the entry of new firms, can earn monopoly profits, control technological progress, and exercise considerable political influence. More important in the present context is that dominant firms can switch their capital resources geographically without competitive pressures to constrain their decisions. Moreover, many of the largest firms that emerged in the United States were conglomerates. Individual businesses were commodities to be bought, sold or stripped of their assets. Firm survival did not depend on success in any one product line. Consequently, if a product was relatively less profitable than expected the assets of the plant producing it could be stripped and put into some other more profitable activity.

Technological change and corporate concentration were therefore complementary. The new technologies enabled the new superfirms to fashion capital-switching and relocation strategies without giving up centralized control. In these circumstances locations outside the historic industrial core could become competitive. The new technologies also made unskilled labor, as long as it was cheap, more, rather than less, attractive. In the years after the war, inside the factory, and later the office, new machinery and new methods of work organization reduced the need for high-priced skilled labor. For example, the assembly-line principle was applied to all sorts of production from food-processing to assembling television sets to packaging cereals. More recently, numerically controlled machine tools and robots have either displaced workers altogether or require low levels of machinery and other skills. The highly skilled labor to be found only in the historic industrial core was no longer needed in many industries (*Business Week* 1986: March 3).

Federal government policies and spending were also important in encouraging shifts in investment from the core to the periphery. They have also encouraged shifts in industrial location from central city to suburban areas in both the core and the periphery (Gottdeiner 1985). At the regional scale Sale (1975), for example, identifies what he calls the 'six pillars of the southern rim' in his discussion of economic growth in the American periphery in the period 1940–67. These are agribusiness, high technology, real estate development, the military, energy and leisure. Each of these has been stimulated in one way or another by federal policies designed to stimulate the national economy through business activity. Agricultural production, for example, has been directly stimulated by government price support programs. This has led many national corporations to diversify into the 'food business.' From 1945 to 1970 the average farm size doubled from 191 acres to 390. In 1969 corporate operations held nearly 40 percent of all acreage in the country, though there are great regional variations. Less land is owned by agribusiness in the Midwest than in California. In California, in 1969, nearly 4 million acres and three-quarters of the prime irrigated land was owned by 45 corporations. The federal government, through programs originally designed to protect small, family farmers from the vagaries of the world market, has ended up sponsoring a modern version of plantation agriculture (Sale 1975).

For a variety of reasons, therefore, including the New Deal coalition, the 'aging' of the industrial core, the complementarity between new permissive technologies and corporate concentration and centralization, government intervention producing originally unintended consequences, and 'Counter-Keynesian' policies favorable to business, the economic dominance of the historic industrial core came into question during the 1950s and 1960s. This was paralleled by shifts in political influence. Some commentators have gone so far as to claim that there had been a 'power shift' in the United States by 1970 from the traditional core to the periphery, especially to the West and Southwest (Sale 1975). But it is only really since then, as will be shown in Chapter 4, that the industrial core has gone into anything like precipitous economic decline. Previously, the decline was gradual and hardly commented on. Only with hindsight can one see the increase in 'regional volatility' that signposted 'trouble ahead.' The signs were political as much as economic.

Beginning as early as 1948, the New Deal coalition showed signs of becoming unglued. The coalition rested upon the continuing tolerance of the northern, working-class faction of the Democratic Party for southern race segregation. In the presidential election of 1948, however, a 'civil war' erupted inside the Democratic Party over civil rights (Lubell 1952: 9). The Dixiecrat rebellion of that year produced a crack in the previously monolithic support given by the South to the Democratic Party. This rebellion echoed through the 1950s and 1960s. There was a 30 percent decline in Democratic Party identification among southern voters (Wayne 1980: 63). By the 1960s Republicans were winning elections in the South. The passage of the 1964 Civil Rights Act was the *coup de grâce*.

During the long period from 1948 until 1964 an uneasy truce of trade-offs and vote-swapping prevailed within the Democratic Party. The seniority and committee systems in Congress encouraged this (Bensel 1984). But there was nothing like the dominance 'core' Republicans exerted in American national politics from 1896 to 1933. Representing core and periphery factions with dissimilar interests produced uneasy compromises rather than singular policies. The national commitment to economic growth allowed the Democratic Party to appeal to all factions. There could be something for everyone. But ultimately this was an unstable interregional alliance founded in the immediate and desperate circumstances of the Great Depression.

The success of the civil rights movement was expected by many of its supporters to change social and political conditions in the South to the extent that the unionization of workers would increase dramatically, the cost of agricultural labor would be raised, and class-based politics such as had prevailed in the industrial core since the New Deal would spread southwards (Bensel 1984). These expectations were to be disappointed. Certainly, the social and political changes of the 1960s hastened the mechanization of southern agriculture and urbanization. But this was at the expense of the industrial North. Vast numbers of poor southerners migrated northwards, supplementing earlier migrants in search of jobs, and when these were not available swelled the numbers of the unemployed and those needing public assistance (see Table 3.6). This increased public expenditures in northern states and required higher levels of state taxation. This in turn discouraged business investment. The southerners who stayed behind neither joined labor unions nor did they participate in class politics. Rather, as Elazar's (1972) argument concerning regional cultures in the United States would lead one to expect, southern white workers inclined to deference towards employers and hostility towards the 'liberal' northern Democratic Party. The 'nationalization' of the South, then, was nothing of the sort. Rather than becoming like the industrial core, the continuing cultural particularity of the South produced a 'climate for business' that was bad for business in the North (Newman 1984; Weinstein *et al*. 1985).

At the same time as southern defection from the New Deal coalition introduced doubts about the Democratic Party's ability to substitute for the old Republican Party as the 'natural party' of national government, the West produced a new Republican Party that portrayed that region as the 'wave of the future.' The origins of this party lay in the conservative reaction to the New Deal. Even if one interprets the New Deal as more a reorganization of power at the top than a social democratic surge from below, it still marked a major watershed in the redistribution of social and economic power in the United States. Eisenhower swept the West in 1952 and again in 1956. Since then the region has been a Republican stronghold, suggesting that 1952 was a watershed or even a realigning election. The Party itself began to take on a western image.

Table 3.6a. *Distribution of black population in the United States by region, 1960–80 (%)*

	1960	1970	1980
New England	1.3	1.7	1.8
Middle Atlantic	14.7	17.5	16.5
East North Central	15.3	17.2	17.2
West North Central	3.0	3.1	3.0
South Atlantic	31.0	28.3	28.9
East South Central	14.3	11.4	10.8
West South Central	14.7	13.3	13.3
Mountain	0.6	0.8	1.0
Pacific	5.1	6.7	7.5
Total	100.0	100.0	100.0

Source: US Bureau of the Census 1981.

Table 3.6b. *Estimated net black intercensal migration by region, 1871–1980*

Intercensal period	South	Northeast	North Central	North Total	West
1871–80	−60	+24	+36	+60	(na)
1881–90	−70	+46	+24	+70	(na)
1891–1900	−168	+105	+63	+168	(na)
1901–10	−170	+95	+56	+151	+20
1911–20	−454	+182	+244	+426	+28
1921–30	−749	+349	+364	+713	+36
1931–40	−347	+171	+128	+299	+49
1941–50	−1,599	+463	+618	+1,081	+339
1951–60	−1,473	+496	+541	+1,037	+293
1961–70	−1,380	+612	+382	+994	+301
1971–5	+14	−64	−52	−116	+102
1976–80	+195	−175	−51	−226	+30

Note: Numbers in thousands. Plus sign (+) denotes net in-migration; minus sign (−) denotes out-migration.
Source: US Bureau of the Census 1981.

Until the 1950s the presidential level of the Republican Party was dominated by its so-called 'eastern establishment' or, as conservatives have called it since 1940, the 'Liberal eastern establishment.' The historic power of this wing lay in industrial and finance capital in the Northeast. Investment banks, in particular, such as the House of Morgan, were central within the Party. These banks had been the channels of the European capital that helped finance the US industrial revolution in the nineteenth century and they used their power to restructure the American economy between 1898 and 1902. By the 1930s, however, industrial corporations and commercial banks had displaced the investment banks from their key position. These elements tolerated the New Deal and were major partners in the alliance between government and business after World War II.

As early as 1938 an 'alternative Republicanism' emerged to challenge both the New Deal and liberal–internationalist Republicanism. This was a peripheral capitalist bloc of banking and industrial interests in the West, but especially strong in the small cities of the Midwest (Illinois, Indiana, Ohio, etc.). Its distinctiveness lay in a blending of sentiment hostile to the New Deal with fierce nationalism – mistakenly labelled 'isolationism' – that urged American expansion in Asia and rejected involvement in European 'quarrels.' Its leader, Senator Taft of Ohio, gained control of the Republican National Committee, but the presidential nomination was always denied him by the 'eastern' wing.

Only between 1958 and 1963 was the 'old guard' resurrected as the 'New Right' (Davis 1981). This time the base was a new peripheral business group from the Far West, especially California. It shared the traditional right-wing interest in the Pacific, but it was much more of a *movement*. It focused upon disaffection with social change and the growing backlash against the civil rights movement. Single issues such as gun control, abortion, open housing, anti-poverty measures, Cuba, homosexuality and women's rights became its rallying points. It was opposed to them. Essential to the orchestration of these single issues was the availability of plebiscites around which public attention could be focused with large inputs of money and advertising. California provided these. It also provided an unstable and uncertain cultural environment in which people could be easily persuaded that their social worlds and the values they held dear were disintegrating. Elazar's (1972) map of political cultures within the states shows California as the only one with all three of the distinctive cultures in volatile combinations (see Figure 3.9) (see also Wilson 1975).

Ultimately, however, western Republicanism also appealed to fundamental economic interests. Issues vital to the West, such as scarce water, government-owned land and illegal immigration from Mexico, matter little elsewhere in the United States. Moreover, the West was no longer by the 1950s the creature of the industrial core (Worster 1986). It had generated its own capitalist class specializing in computers, semiconductors, aerospace and real estate. Combining high-tech industries with high-tech agriculture and a vast array of service-related industries, the Pacific Coast states have acquired a booming trade with Asia that outstrips exports elsewhere. In the final analysis, however, western Republicanism is an ideology and movement the historian of the frontier, Turner, would recognize. In 1932, just before the coming of the New Deal, Turner (1932: 254) observed of the West: 'This land of farm owners, this land trained in pioneer ideals, has a deep conservatism, at bottom, in spite of its social and political pioneering.' Most of the farm owners have gone; the conservatism remains.

From World War II until the late 1960s, then, the clear dominance of the northeastern industrial core within the American political economy declined. After the war the Democratic Party sacrificed control over the economy in return for an expansion of it. The main benefits of the growth that did follow did not accrue to the industrial core, as a whole, which actually began to decline, but to the western periphery sectors of high technology, military hardware and agribusiness, the southern periphery sector of labour-intensive manufacturing, and high-income/low-tax suburban areas. Unwittingly, the macroeconomic approach put in place by the Democratic Party created the conditions for its own demise as a *national* party.

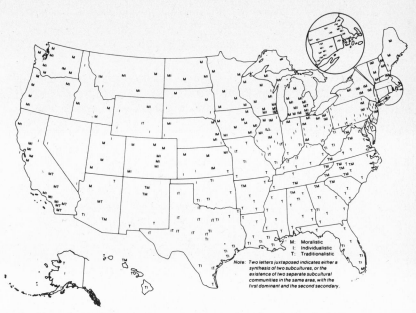

Figure 3.9: The distribution of 'regional' political cultures within the states (Elazar 1972: 124). Reproduced by permission of Harper and Row, Publishers, Inc.

But the period 1940–67 was still one of general growth. Decline was still only relative. Most major national corporations were still head-quartered in the northeastern core – even if their physical operations were no longer in the manufacturing belt. Across a wide range of indi-cators the standard of living or level of social well-being in 1970 was also still considerably higher in the industrial core than in the peripheries, especially the southern one (see Figure 3.10). Regional volatility had not led to a reversal in the roles of core and periphery. The core was slipping, but it was not yet eclipsed.

Conclusion

The history of the United States in the world-economy is a history of regions relating to that world-economy in fundamentally different ways. From the colonial period until the Civil War there was marked competition within the United States between regions with remark-ably different relationships to the world-economy. The industrializing North was victorious, and as the United States became predominant in the world-economy after 1896 the sources of regional difference were more and more generated within the boundaries of the country itself. The industrial core of the United States was the dominant element in the core of the world-economy. Consequently, a geographical division

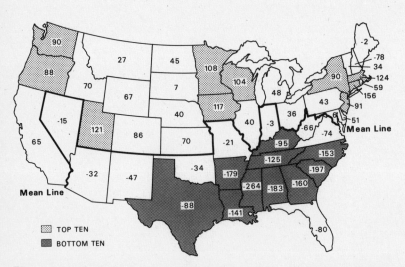

Figure 3.10: State scores on a composite indicator of social well-being (Smith, D.E. 1979: 11). Reproduced by permission of Penguin Books and the Johns Hopkins University Press (US)

of labor between the economically advanced northern core and the underdeveloped southern and western peripheries reproduced *within* the United States the fundamental division within the world-economy. Since World War II this pattern of regional dominance/dependence has become less marked. However, at least in the late 1960s, a new regional pattern had not yet appeared.

4

The American impasse and America's regions

The period 1967–86 has seen a major downturn in the world-economy with major effects on the United States. But it is not some *Geist* or superorganic force that caused the problems now faced by the United States within the world-economy. Rather, the downturn is itself the product of *decisions* made within the United States, the dominant entity within the world-economy (see Chapter 1). The tremendous economic growth experienced in the 1950s and 1960s was achieved at long-term cost. The basis for this growth – economic concentration, government intervention and overseas expansion – also created a set of conditions that eventually limited growth. In particular, economic concentration led to the growing globalization of American business, and overseas expansion led to increased defense expenditures and a drain on the US balance of payments. The problems of 1967–86 are a direct inheritance of decisions made and put into practice in the period 1945–50, and earlier.

The purpose of this chapter is to identify the major elements of the impasse the United States now faces within the world-economy and relate these to the changing internal geography of the country. The chapter is divided into four sections. In the first section attention focuses on the consequences for the American 'standard of living' of the recent decline in growth. The second section attempts to detail the various causes of this decline. In the third section general regional impacts are identified. A fourth, and final, section presents a series of vignettes or case studies describing specific local impacts in a selection of American cities and regions.

130

The American standard of living

In the twenty years after World War II the United States was clearly the world's richest country. While much of the industrialized world was clearing the rubble from its bombed-out cities and devastated factories, the United States was capturing resources and building markets that translated into unprecedented affluence for much of its population. In the 1970s, however, the country witnessed a precipitous decline in its ability to create affluence. After twenty years of growth came stagnation and decline. Between 1970 and 1979 the real median family income of Americans increased by only 6.7 percent. In 1980 it actually decreased by 5.5 percent. This compared to an increase of 37.6 percent in the 1950s and 33.9 percent in the 1960s (Magaziner and Reich 1982).

The growth in income per employed person was also considerably lower in the United States than in other industrialized countries in the period 1960–78 (see Figure 4.1). By 1979 nine other industrialized

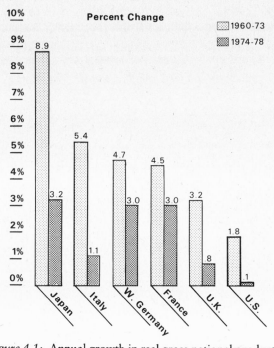

Figure 4.1: Annual growth in real gross national product per employed person (Franko 1980). Reproduced by permission of the author

Table 4.1. *Gross domestic product per capita as a percentage of US GDP per capita, 1960–79*

	1960	1963	1970	1975	1979
Switzerland	57	66	70	118	139
Denmark	46	53	67	104	119
Sweden	67	75	86	118	115
Germany	46	53	64	95	116
Iceland	49	56	51	82	103
Norway	45	50	60	99	106
Belgium	44	47	55	90	107
Luxembourg	59	57	66	89	109
Netherlands	47	55	58	90	101
France	47	55	58	90	100
Canada	79	72	81	101	91
Japan	16	22	41	63	82
Finland	40	45	48	82	82
United Kingdom	48	50	46	58	67

Source: OECD 1961–81.

countries had higher absolute income levels than the United States (see Table 4.1).

Relative and absolute income figures tell only part of the story. Such information gives only a rough quantitative estimate of the goods and services that can be consumed by a national population. It does not reflect such important but less quantifiable aspects of a 'standard of living' as health care and longevity, availability of leisure time, a clean environment and personal safety. Neither does it indicate anything about the distribution of income and wealth. However, in these respects as well, the United States now compares unfavorably to other countries. By way of example, workers in the United States stand a much greater chance of becoming unemployed without adequate insurance than in other industrial countries (see Figure 4.2). Moreover, most workers in the United States can be dismissed summarily without notice or reason; in other industrial countries this is not the case. Without work Americans find that there is limited public expenditure for social welfare. Most industrial countries outspend the United States considerably on social expenditures (see Figure 4.3). If anything this difference has become even more pronounced since 1979 (*Focus* 1984).

Workers in the United States also have much less leisure time than workers in Europe and Japan (Magaziner and Reich 1982: 17). In the United States the average annual vacation is only 2½ weeks compared to *minima* of 4 weeks in Japan and most European countries.

The same story is repeated across a range of quality of life indicators. For example, the United States ranks poorly in life expectancy and infant mortality – important and reliable measures of public health. In 1975 fourteen countries ranked ahead of the United States in male life expectancy. Seven of them exceeded the United States in female life expectancy (United Nations 1978: 402–14). The United States declined from fifth place in infant mortality in 1950 to eighteenth place in 1977 as most industrial countries, and even a number of underdeveloped ones, exceeded it. The United States also does poorly when compared to other countries in terms of such measures of life quality as pollution and the likelihood of suffering from violent crimes (Magaziner and Reich 1982: 21–2).

Finally, despite all the talk in the United States about anti-poverty programs and 'big government' reducing individual incentives for work by overrewarding the poor and the spendthrift, the United States has a remarkably inegalitarian distribution of income. Of industrialized countries only France and Canada provide an equally low share of national income to their poorest people (see Figure 4.4). Moreover, the gap between the richest and the poorest in the United States has widened, especially since 1960 (Magaziner and Reich 1982: 24).

These comparisons are not intended to suggest that the United States is, at least in a quantitative sense, necessarily a worse place to live in than the other countries, for whether it is or not depends on what sort of job, what sort of income security, what sort of health insurance, and what sort of place you live in. Rather, they are presented to show that there has been a dramatic change in the relative American standard of living. This should rebut the idea that the problems of the 1970s and 1980s are merely temporary difficulties and the United States is still 'far ahead' of other countries. It should also challenge the idea that some vast redistribution of income and wealth in the United States has undermined investment and productivity, as alleged by many conservative politicians (Edsall 1984). The declines in investment, productivity *and* the relative standard of living in the United States have quite different origins.

Elements of the American impasse

What went wrong in the 1960s? What made the US economy slip off the growth track it had been on since World War II? Numerous theories have been put forward to account for this. One implicates the growth of government spending for social programs and the introduc-

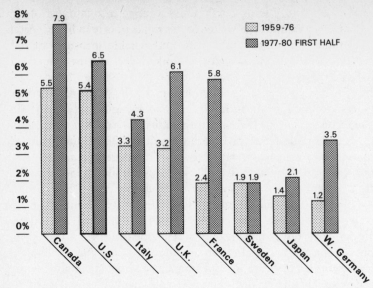

Figure 4.2a: Average unemployment rates for selected countries (Magaziner and Reich 1982: 14–15)

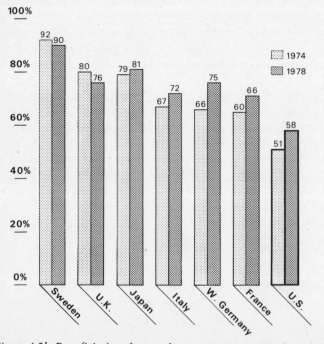

Figure 4.2b: Beneficiaries of unemployment insurance in selected countries as a percentage of total people unemployed (Magaziner and Reich 1982: 14–15)

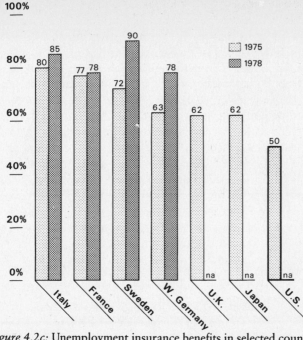

Figure 4.2c: Unemployment insurance benefits in selected countries as a percentage of average earnings (family of four) (Magaziner and Reich 1982: 14–15)

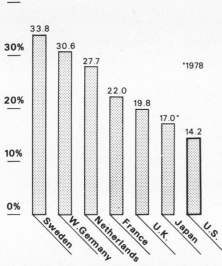

Figure 4.3: Public collectivized expenditure on social welfare (health care, welfare assistance, unemployment benefits, etc.) as percentage of GNP for selected countries, 1977 (Magaziner and Reich 1982: 16)

tion of pollution controls as measures which have discouraged investment. Another implicates government deficits as the sole culprit,
blaming them for 'crowding out' private borrowers and thus crimping
investment in new plant and new industries in the United States.
Others suggest low labor mobility and strikes as the villains of the
piece. But in fact US labor mobility is much higher than that in Japan
(Fieleke 1981b) and strikes, particularly long ones, have been on the
decline in number and impact throughout the 1970s and early 1980s
(Noble 1985). Finally, some commentators complain that Americans
'don't want to work anymore' and are idle on the job, or that changes
in the composition of the work force have led to declining productivity. It is clear, however, that in countries with considerably less
managerial discipline in the workplace than the United States has
productivity is much higher (e.g. Sweden, Denmark and Holland).
Furthermore, studies have shown that the entry of women and
minority groups into manufacturing industries in which they were
previously absent is a *consequence* of these industries' inability to
generate higher-value jobs rather than a *cause* of lower productivity
(Denison 1979).

What these approaches share is a single-minded focus on the
national scale. They ignore the historical evolution of the United
States within the world-economy. Most also offer an easy and, to their
proposers at least, painless solution: remove government social and

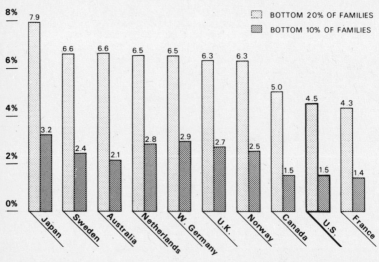

Figure 4.4: The distribution of after-tax income shares, 1975 (Magaziner and
Reich 1982: 23)

environmental programs and the golden age of growth will return. However, as the data in the previous section strongly suggest, there is no simple inverse correlation between government social spending or government intervention in the economy *per se* and economic growth.

A more sophisticated explanation would focus on the international context for the economic shift. Though there might be differences of emphasis, a general synthesis would stress in the first place (1) the significant decline in US relative productivity since 1965. It would then turn to two trends that correlate with the productivity decline: (2) the passing of technological leadership, especially in non-defense research and development, from the United States to other countries and (3) the increased 'openness' of the US economy to foreign competition. But what, one might ask, has brought about the three trends? It is at this level that the 'globalizing' effects of the growth coalition identified earlier appear important. They include (4) the increased investment of US-based multinational enterprises (MNEs) abroad; (5) increased competition for these firms from MNEs based elsewhere, many of which owe their origins to previous American direct investment; (6) the increased dependence of the US economy on foreign sources of raw materials as a result of depletion at home and vast investments abroad; (7) the drain on the US economy of the *Pax Americana*: defense spending and the costs of military intervention in Vietnam and elsewhere; and (8) the priorities of domestic 'Counter-Keynesianism,' particularly the emphasis on housing as an economic 'stabilizer,' have diverted capital away from investment in new plant and new industries (see Figure 4.5). Economic concentration, government inter-

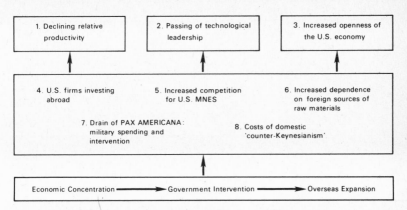

Figure 4.5: Elements of the American impasse in the world-economy

vention and overseas expansion, then, have jointly produced decline where they once produced growth.

Each of the elements of the 'American impasse' is now examined in turn.

Declining relative productivity

The relative position of the US economy within the world-economy is best measured by relative productivity. Productivity refers to the value-added per work hour in producing goods and services. It can be increased in two ways, by improving the value-added in existing industrial activities or by shifting resources to activities that produce relatively more value-added than existing ones. In both ways the US economy has been failing since the late 1960s. Productivity growth has declined markedly over the past twenty years. Measured as total output in the economy divided by the number of hours worked, productivity increased by an average of 3.2 percent per year from 1948 until 1965. It fell to 2.4 percent between 1965 and 1973 and to 1.1 percent between 1973 and 1978. Productivity was negative from 1978 to 1980: −0.8 percent in both 1979 and 1980. Since then it has rebounded somewhat to around 1.2 percent by 1984 (Sadler 1982; *Business Week* 1984: February 13). As yet, however, there are no signs of the great 'Revival of Productivity' predicted by *Business Week* magazine (1984: February 13) during the cyclical upturn of 1983–4 (US Department of Commerce 1985).

More importantly, from the perspective of the US economy within the world-economy, the United States has lagged behind other countries in productivity improvements for the past twenty years (see Figure 4.6). The United States started the 1950s with much higher levels of absolute output per worker, but the slow rate of growth in productivity has allowed many others to surpass it in many industries. Lower productivity has in turn meant lower increases in incomes (see Figure 4.7).

Passing of technological leadership

One explanation for the more rapid growth of productivity in other countries relative to the United States is that they have been 'catching up' with the United States by using technologies already developed by the United States. This was true of Europe and Japan in the 1950s and 1960s, and Japan in the early 1970s, but it is no longer valid. Manufacturers in other countries have now achieved technological leadership across a wide range of products in design, quality and productivity. One measure of this change is the US patent balance with other countries (see Table 4.2). This has become less and less favorable to the United States. In particular, the US patent balance with West Germany turned negative in the mid-1960s and with Japan in the mid-1970s. The United States is now more likely to be importing Japanese and German innovative technologies than exporting its own to those countries.

Another measure of technological leadership is the proportion of GNP spent on research and development. This has been declining in the United States, while that of other countries has been increasing rapidly (see Table 4.3). Of course, absolute levels of spending are still much higher in the United States given the larger size of the US GNP. But much of this is spending on military research that has limited commercial stimulus. For example, in three key industries – electronics, aircraft and machine tools – firms serving as prime defense contractors have *lost* rather than gained market share (Council of Economic Advisors 1981; Haavind 1983; Schlosstein 1984). They have lost most of this to foreign competitors.

Another plausible contributing factor to the decline of American technological leadership is the increasing conglomerate form of American corporate organization. For example, in the period 1976–9

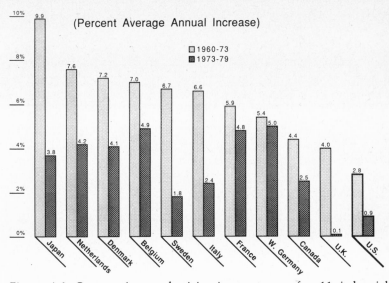

Figure 4.6: Comparative productivity improvements for 11 industrial countries, 1960–73 and 1973–9 (Magaziner and Reich 1982: 36)

the US Steel Corporation reduced its capital expenditures in steel-making by a fifth. Profits were instead directed into the acquisition of chemical firms, shopping malls and other activities. In 1979 US corporations made acquisitions totalling $40 billion, more than the total spent on research and development by all private firms in the country (*Business Week* 1980: June 30). Not only are funds that could

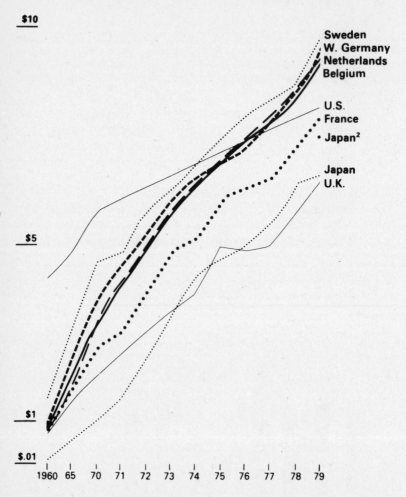

Figure 4.7: Estimated hourly compensation[1] for production workers in manufacturing (includes benefits), 1960–79[2] (Magaziner and Reich 1982: 37)

[1] Annual exchange rate; includes regular benefits (pension, insurance, etc.).
[2] 1979 figure includes extra benefits received by Japanese workers (e.g. low interest loans).

Table 4.2. *The US patent balance (number of patents granted to individuals and companies), 1966–75*

	1966	1972	1975
West Germany			
Granted to US	3,733	4,575	3,140
Granted by US	3,981	5,728	6,069
Balance	−248	−1,153	−2,929
Japan			
Granted to US	4,683	5,948	4,918
Granted by US	1,122	5,154	6,339
Balance	3,561	794	−1,421
Total			
Granted to US	45,633	30,520	37,482
Granted by US	9,567	16,839	19,197
Balance	36,066	13,681	18,285

Source: Magaziner and Reich 1982: 52.

Table 4.3. *Research and development as a percentage of GNP for selected countries, 1962–78*

	1962	1964	1966	1968	1970	1972	1974	1976	1978
United States	2.73	2.97	2.97	2.83	2.64	2.43	2.32	2.27	2.25
Japan	1.47	1.48	1.48	1.61	1.79	1.85	1.95	1.94	1.98
West Germany	1.25	1.57	1.81	1.97	2.18	2.33	2.26	2.28	2.28
France	1.46	1.81	2.03	2.08	1.91	1.86	1.81	1.78	1.85
United Kingdom	na	2.30	2.32	2.29	na	2.06	2.00	1.20[a]	2.10[a]

[a] Estimated.
Source: Magaziner and Reich 1982: 53.

go into the search for new technologies thus diverted, the orientation of a firm towards a single industry or product line shifts from commitment towards developing it to concern with its present contribution to company profits.

Clearly, technical innovation plays an important role in fostering increased productivity and stimulating long-term economic growth (McKinney and Rowley 1985). It has been the major driving force behind the long cycles of the world-economy. Just as in the late nineteenth century the European countries, especially Britain, lost their predominance in technical innovation to the United States, so today the United States is losing its to Japan and other countries. In one key industry, semiconductors, the US industry is now well and truly eclipsed by Japanese firms (Sanger 1985). This will greatly harm the competitiveness of the US electronics industry in general (Semiconductor Industry Association 1983).

Increased openness of the US economy

Prior to 1970 only a small fraction of US manufacturing industry was exposed to world trade. American firms needed only to be concerned about their productivity relative to other American firms. That is no longer the case. In 1980 17 percent of goods produced in the United States were exported (9 percent in 1970) and over 21 percent of goods sold in the United States were imported (9 percent in 1970). Almost 25 percent of the growth in US consumption of goods over the period 1970–80 was taken by imports. Even this incredible rise in imports and exports, however, underestimates the new importance of foreign competition. Although 25 percent of the steel used in the United States is imported (15 percent in 1970), total steel production is threatened by foreign competition. If one adds up the total production of US goods in Standard Industrial Classification six-digit categories where imports equal 10 percent or more of consumption and exports equal 10 percent or more of production, 75 percent of American goods must now face competition in an international market. One industry in which the foreign share of the American market has risen most dramatically is machine tools. Whereas in 1975 9.5 percent of the machine tools sold in the United States were imported, by 1980 this had risen to 23.6 percent. In 1984 it stood at 41.5 percent (Farnsworth 1984).

In this setting, relative productivity declines, such as those noted earlier, become *cumulative*. If US productivity falls relative to that of other countries producing the same internationally traded goods, then either output decreases and unemployment increases, or US workers must accept lower relative incomes, or both. An important feature of international trade in the 1970s and 1980s has been the growth of 'intra-industry trade,' the simultaneous export and import by a country of products in the same industry. It is also called 'trade overlap' or 'two-way trade' (Bergstrand 1982). This reflects both the growth of differentiated products within industries and the return to comparative advantage through specialization and deliberate attempts to capture foreign markets, whether or not a particular country (e.g. Japan) has a 'natural' comparative advantage. One by-product of this trade, however, is increasing competition *within* industries *between* countries. In this context even a short-run decline in relative productivity due to faulty management or failure to innovate can produce dramatic long-run declines in market share.

The United States has suffered a major decline in its ability to be as productive as other countries that produce the same goods. Productivity, therefore, is not simply relative, it is also competitive. Domestic production has been displaced by imports in an increasing number of goods. In 1979 the United States imported 21 percent of its cars, 16 percent of its steel, 50 percent of its televisions, radios, tape recorders and record players, and 90 percent of its knives and forks. US exports have not kept up; the US balance of trade has become increasingly negative since 1970 (Magaziner and Reich 1982: 31–5). This happened during both a period when a lower-valued dollar should have made American exports cheaper and imports more expensive (1971–8) and when an overvalued dollar would lead one to expect a negative trade balance (1978–85).

In addition to the increased level of trade, there have been changes in the nature of trade. The United States now exports proportionately less manufactured goods and exports far more agricultural products, high-technology devices and business services. Much greater percentages of manufactured goods are imported, especially those that result from labor-intensive processes. In 1984 the United States had a $100 *billion* deficit in trade, mainly in manufactured goods and petroleum products. There was a large surplus in business and other services.

The service balance makes the US current account consistently better than the trade balance would lead one to expect. The greatest portion of the positive US service balance derives from the investment incomes of American manufacturing firms and, to a much lesser extent, individuals. However, the ratio of receipts to payments has declined significantly since the 1960s. In 1955 the United States received five times as much income as it paid out. By 1978 the ratio of receipts to payments had shrunk to 1.8 : 1. Fees and royalties from subsidiaries abroad constitute the second major component of the service balance. This balance has also declined rapidly as American manufacturing firms have lost their competitive edge. Finally, service exports such as banking and insurance form the third component of the service balance. These services have not declined to anything like the same extent. But ultimately they, too, depend on the maintenance of a healthy manufacturing base. As British banks know only too well, there is no true comparative advantage in service exports without a manufacturing base to build on (Magaziner and Reich 1982: 85–6).

Table 4.4. *The direction of trade for the United States for selected world regions and countries (%), 1970–8*

	1970		1978	
	Exports	Imports	Exports	Imports
Regions				
World	100.0	100.0	100.0	100.0
Industrial countries	61.2	68.0	53.2	52.8
Other Europe	4.3	2.5	3.6	2.0
Australia, New Zealand,				
South Africa	3.9	2.8	3.1	2.6
Oil-exporting countries	5.0	4.8	11.8	17.7
Other less developed areas	25.1	21.3	25.6	24.0
USSR, East Europe, China, etc.	0.7	0.5	2.9	0.9
Countries not specified	0.0	0.5	0.1	0.0
Selected countries				
United Kingdom	5.9	5.5	5.3	3.8
Japan	10.8	14.2	9.0	14.4
France	3.4	2.4	2.9	2.4
West Germany	6.3	7.8	4.8	5.8
Ireland	0.3	0.3	0.4	0.2
Portugal	0.3	0.2	0.4	0.1
Greece	0.5	0.1	0.5	0.0
Turkey	0.7	0.2	0.2	0.1
Mexico	3.9	3.1	4.6	3.4
Brazil	1.9	1.7	2.1	1.7
Republic of Korea	1.5	0.9	2.2	2.2
Republic of China	1.2	1.4	1.6	3.1

Source: International Monetary Fund 1979.

The direction of trade has changed as well. The United States now imports relatively more goods from less developed countries and the so-called 'newly industrializing countries' (Brazil, Mexico, South Korea, Singapore, Taiwan, Hong Kong) and less from other industrial countries (see Table 4.4). The increase of US imports from the less developed countries is based primarily on lower costs, which are in turn due to their low wage rates.

The growing involvement of the United States in world trade has been facilitated by a number of factors. One, already mentioned, is the lower wage rates and lower costs of producing many goods outside than inside the United States. A second factor is technological change in communication and transportation which has reduced the costs of decentralized production activity (noted in Chapter 3). Associated with this have been technical changes that reduce the need for skilled labor and thus make low-wage production sites more attractive.

Third, there is the important role of multinational enterprises (MNEs) in organizing and controlling international trade. MNEs have increased the flow of capital and technology around the world. Developing countries now have access to capital and technology that they were previously denied. This poses two problems for the US economy: capital and technology from the United States are no longer 'captive' within the United States and capital export on a massive enough scale can lead, when the world-economy slows down, to repayment crises, such as those facing many less developed countries, and to a subsequent decline in exports from the United States as affected governments restrict their imports (Farnsworth 1983).

Fourth, the growth of large-scale retail outlets in the United States (Sears, K-Mart, J. C. Penney, etc.) has enabled products from developing countries to be sold at competitive prices in the United States. A Korean television manufacturer, for example, can gain a share of the American market by selling to only three or four large US department store chains.

Fifth, and finally, since 1980 foreign producers have been given a sharp competitive edge by the huge fiscal stimulus to US consumption provided by the ballooning deficit of the US federal government. The US government's financing needs boost interest rates, which in turn bolster the dollar. This makes imports relatively cheaper and US exports relatively more expensive. For the sake of increasing consumption and defense spending at home the US government has been stimulating business abroad relative to that in the United States.

The threat to American economic growth from trading activity, therefore, is broadly based. On the one hand, labor-intensive industries are increasingly threatened by competition from the less developed countries. On the other hand, more technologically complex industries have serious trade imbalances with Japan and the European Economic Community. These latter industries constitute the US export base, so import erosion of the domestic market is especially serious. Of current exports 75 percent are in this category. As competitive imports increase, can a decline in exports of like products be far behind?

Throughout the period 1967–86 American governments have remained committed to the idea of 'free trade.' Although numerous 'voluntary quotas' and restrictive trade practices have been introduced, there has been only a limited embrace of protectionism by powerful political and business leaders. This marks both the extent of the globalization of American business and the continuing commitment to the idea of identity between the US economy and the world-

Table 4.5. *Outward direct investment flows for 13 OECD countries,*
1961–78 (% distribution)

	1961–7	1968–73	1974–8
Canada	2.3	4.5	6.2
United States	61.1	45.8	29.6
Japan	2.4	6.7	13.2
Australia	0.7	1.4	1.6[a]
Belgium	0.3[b]	1.4	2.1[a]
France	6.9	5.2	7.8
Germany	7.2	12.5	16.7
Italy	3.6	3.3	1.9
Netherlands	4.4	6.8	9.8
Sweden	2.0	2.4	3.7[c]
United Kingdom	8.7	9.1	7.9
Spain	0	0.3	0.6
Norway	0	0.3	1.0

Note: The UK and US figures do not include reinvested earnings to place the data on a
more comparable basis. UK data do not include the petroleum sector.
[a] From 1974 to 1976.
[b] From 1965.
[c] From 1974 to 1977.
Source: OECD 1980.

economy. Yet many of America's leading trading 'partners,' especially
Japan, operate according to very different principles. Trade is viewed
by them in mercantilist rather than free-trade terms. They see their
international business as a means to serve national ends rather than
the goals of individual corporations.

Increased investment by US firms abroad

Underpinning the decline in relative productivity, the loss of technological leadership and the increasing openness of the US economy to
foreign competition are a number of economic–institutional factors.
One of these is continuing and increasing investment abroad by US
firms. As they have been disinvesting in domestic industry large US
firms have continued to shift capital beyond the country's borders.

There have been major shifts in the direction of US direct foreign
investment in the 1970s. The proportion of investment in other
industrial countries peaked in 1964 (see Table 4.5). Thereafter there
was increasing investment in less developed countries. There has also
been a change in the nature of American direct foreign investment (see
Figure 4.8). In particular, there has been an increasing proportion of

foreign investment generated by the reinvestment of retained earnings from subsidiaries (internal cash flow). This reflects the fact that profits are not taxable unless they are returned to the United States.

The incentives for investment abroad have remained much the same as those from an earlier era. At least through the 1970s foreign locations provided numerous advantages: lower wage rates, access to protected markets, the absence of workplace and environmental regulations, and tax advantages. Even with the increased attention given to their needs after the 1980 presidential election, MNEs remain firmly

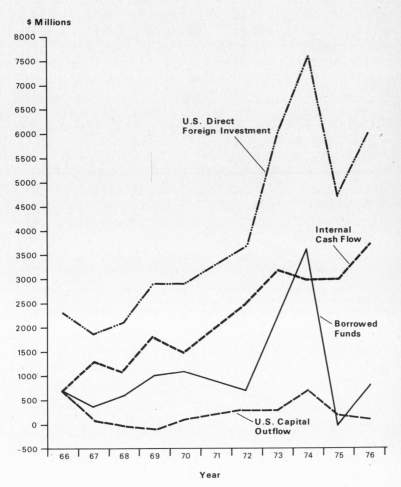

Figure 4.8: Means of financing US direct foreign investment, 1966–76 (OECD 1980)

wedded to the global strategy: spreading risks geographically and constantly searching for the most favorable profit opportunities.

Some MNEs are now in the process of moving all their manufacturing operations abroad or ceding production to foreign companies for whom they then act as marketing agents (*Business Week* 1984: October 8). A major example is in the automobile industry, where General Motors is implementing its so-called 'Asian strategy': the company is selling small cars made by Japan's Suzuki Motor Company and Isuzu Motors, and by 1987 it will begin marketing small cars produced in South Korea by Daewoo Corporation. Joint ventures such as these are part of a strategy to slash costs by buying cheaper, foreign-made components, or entire products whenever possible (*Business Week* 1986: March 3).

The strength of the dollar relative to other currencies since 1978 has encouraged MNEs to look abroad for joint-venture partners and for manufacturing capacity. Otherwise their products would become uncompetitive in the foreign markets they have dominated for years. The long-term cost is the sharing of control, and profits, that joint ventures and overseas 'sourcing' of components involve.

For the US economy the major consequences are loss of capital for domestic investment, loss of jobs and loss of technological advantage. The rate of capital formation in the United States has been consistently lower than that in most other industrial countries since the 1960s (Magaziner and Reich 1982). This reflects both the low rate of saving in the United States – around 4 percent of disposable personal income in recent years, actually 2.5 percent in 1985 (Hershey 1983), compared to Britain's 14.5 percent and Japan's 20 percent – and the 'maturity' of many of its industries. However, of what was invested in manufacturing one-tenth or more was invested by US companies *outside* the United States. In 1979, for example, US companies invested 9.5 percent of their total plant and equipment expenditures abroad, compared to 2.7 percent for West Germany and France and 1.9 percent for Japan (Magaziner and Reich 1982: 48).

US companies have also been aggressive in moving their factories overseas to reduce labor costs. In 1972 42 percent of US imports came from such operations (Barnet and Müller 1974: 266). This is particularly true of the clothing, consumer electronics and small-appliance industries. They often move as an alternative to investment in new technology at home. Japanese and German companies, however, have largely avoided this approach. They have favored automation and other technical solutions to the low-wage solution of American companies. The long-term benefits of the former strategy in terms of increased relative productivity are obvious.

In addition, American companies not expanding abroad themselves but with their eyes set on short-term profits have licensed technologies to foreign firms without any attention to the long-term consequences. Licensing technology is an easy way of paying off development costs and making quick profits without additional capital investment. But it is also dangerous. Licensees have often become competitors. With this in mind, German and Japanese companies have been much more conservative in their licensing practices than have American ones. This has been much to their advantage.

US banks have also been active investors abroad. In the 1970s the eight largest American banks lent more than 100 percent of their shareholders' equity to companies (including the subsidiaries of American MNEs and governments in four countries – Mexico, Brazil, Venezuela and Argentina (Thurow 1984). If these four countries were to default on their payments, the banks would instantly have negative net equity. But in the 1970s the 'earnings picture' outside the United States was much rosier than inside and there were no official constraints on foreign lending, so the big banks contributed their share to capital flight from the United States (see Figure 4.9).

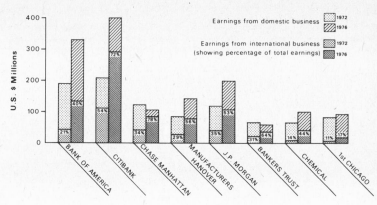

Figure 4.9: International and domestic earnings of eight large American banks, 1972 and 1976 (OECD 1981)

Table 4.6. *Distribution of the world's 50 largest corporations by year and headquarter country, 1956–80*

	United States	Europe	Japan	Iran, Brazil, Mexico, Venezuela
1956	42	8	0	0
1957	32	7	0	0
1958	44	6	0	0
1959	44	6	0	0
1960	42	8	0	0
1961	40	10	0	0
1962	39	11	0	0
1963	39	11	0	0
1964	37	13	0	0
1965	38	12	0	0
1966	38	12	0	0
1967	39	11	0	0
1968	37	12	1	0
1969	37	12	1	0
1970	32	13	5	0
1971	30	15	5	0
1972	27	17	6	0
1973	24	20	6	0
1974	24	20	4	2
1975	23	20	5	2
1976	22	20	5	3
1977	22	19	6	3
1978	21	20	6	3
1979	22	20	6	2
1980	23	20	5	2

Source: Bergesen and Sahoo 1985: 599.

Increased competition from foreign MNEs

American dominance of the world's largest firms has declined dramatically since the late 1960s (see Table 4.6). In 1956, 42 of the world's top 50 firms were American. By 1980 this number had dropped to 23. Europe as a whole now almost equals the United States. Japan's appearance on the list occurred suddenly in 1970 when it acquired 5 firms in the top 50. The turning point was 1970–3, when the United States lost 13 firms from the list. The rise of Japan is noteworthy if only because the *type* of MNEs involved is different from the American. The American MNEs are private capitalist firms whereas the Japanese MNEs are a hybrid between the private firm and the state corporation. Indeed, many of the European MNEs also have 'special relationships' with their respective governments and the European

Table 4.7. *Inward direct investment flows for 13 OECD countries,*
1961–78 (% distribution)

	1961–7	1968–73	1974–8
Canada	16.2	12.1	3.0
United States	2.6	11.4	24.5
Japan	2.0	1.7	1.2
Australia	15.6	12.9	9.5[a]
Belgium	4.5[b]	6.1	9.8[a]
France	8.2	8.2	15.4
Germany	21.3	16.4	17.1
Italy	11.5	8.3	5.9
Netherlands	4.7	8.5	6.3
Sweden	2.4	1.7	9.5[c]
United Kingdom	9.7	7.4	7.6
Spain	2.7	3.7	3.1[c]
Norway	0.8	1.4	4.3

Note: The UK and US figures do not include reinvested earnings to place the data on a
more comparable basis. UK data do not include the petroleum sector, banking and
insurance.
[a] From 1974 to 1976.
[b] From 1965.
[c] From 1974 to 1977.
Source: OECD 1980.

Economic Community. The decline of the American MNEs, therefore, is not only a sign of American decline within the world-economy;
it also represents the decline of a form of production pioneered in the
United States at the turn of the last century (Bergesen and Sahoo
1985).

As a function, in part, of the increased importance of non-American
MNEs in the world-economy the United States had become an important importer of capital from abroad in the 1970s and 1980s. In the
early 1960s only 2.6 percent of the direct foreign investment of the 13
largest OECD (industrial) countries was in the United States. By 1978
this had risen to 24.5 percent (see Table 4.7). Foreign investment in the
United States is now as pervasive as American investment in Europe
was in the 1950s and 1960s. Much of this investment makes up for
that lost through capital export by American MNEs.

The threat and reality of American protectionism (tariffs, quotas,
etc.) have attracted much of this investment as a means of protecting
market shares. Investment has come in some cases to substitute for
exports. This is undoubtedly beneficial in many respects for the US
economy. It can create jobs and through the introduction of new tech-

nology and management techniques can improve relative productivity (*Business Week* 1984: June 4). The problem is that in practice foreign MNEs are primarily engaged in the acquisition of already existing American companies rather than in the creation of new plants and new jobs (Bluestone and Harrison 1982: 157). Moreover, since 1980 the bulk of foreign investment in the United States has been portfolio rather than direct. Investment has been attracted by high interest rates rather than by industrial investment opportunities. Indeed, the over-valued dollar produced by the high interest rates makes foreign *direct* investment in the United States less profitable than investment in other industrial countries. In 1980 average US hourly compensation (earn-ings and benefits) in manufacturing was $11.44 – seventh among 12 industrial countries. By 1983, despite the fact that US compensation had risen only 4 percent in real terms, the rate of $14.14 per hour put the United States at eleventh and highest among the countries sur-veyed. The appreciation of the dollar since then has widened the gap still more. Foreign MNEs, therefore, can 'buy' workers for much less in industrial countries *other* than the United States (*Business Week* 1984: October 15).

The real importance of the foreign MNEs, however, lies outside the United States. They symbolize both the globalization of the world-economy and the rise to dominance of non-American elements within it. American MNEs are increasingly dependent on joint-venture and other schemes involving the collaboration of foreign MNEs (*Business Week* 1984: March 12; May 28). The balance of power has changed. In the 1950s and 1960s American MNEs actually helped create the foreign MNEs they now must compete or curry favor with. In pursuing their objective of gaining access to overseas markets, American corporations licensed technologies, engaged in joint-production arrangements, and invested in the stock of foreign firms. They thus helped raise productivity abroad relative to that in the United States. This only became apparent in the 1970s with the increased openness of the US economy, which was also stimulated in large part by the activities of American MNEs (see the earlier section on 'Increased Openness of the US Economy').

Bluestone and Harrison (1982) provide a number of examples of how American MNEs helped create their future competitors. One example concerns General Electric, which in 1953 bought stock in Toshiba, then a small Japanese company. By 1970 General Electric (GE) owned the single largest block of shares in that firm. GE also owns 40 percent of a subsidiary, Toshiba Electronics Systems Company Ltd, and has 24 licensing arrangements with both companies to make products GE used to make in the United States – radar, generators, lamps and boilers – and which are still sold internationally under the GE label. Bluestone and Harrison (1982: 143) note one of the additional benefits these arrangements provided to GE: 'In 1969 during a big strike against GE in the United States, Toshiba provided its nominal competitor with crucial electrical and electronic parts, something that the unions in other American plants were able to prevent their employers from doing.'

Increased dependence on foreign sources of raw materials

One of the major advantages the United States used to have within the world-economy was its large and diverse stock of raw materials. But in the 1960s the country became more dependent on foreign sources of basic raw materials. One cause of this was the depletion of readily accessible sources within the United States. For example, at one time the United States was self-sufficient in iron ore, but over a period of years the best and most accessible ores in Minnesota and northern Michigan have been mined. Lower ocean transportation costs have also made foreign sources in Australia, Brazil and Liberia more competitive. But most importantly, with the depletion of readily available sources in the United States, the US firms involved in raw-materials exploitation turned increasingly abroad, where they could obtain large quantities at lower unit costs.

In the 1960s, however, many of the countries in which oil production had been developed on a large scale organized themselves into a cartel to restrict competition and through concerted action raise oil prices (OPEC). The most dramatic price increases came in 1973 but were followed by others throughout the 1970s. There was a twelve-fold increase in the price of oil from 1973 to 1981. This was a tremendous shock to all industrial nations, which had taken cheap energy sources for granted. But it was especially damaging to the US economy, which had been built in the 1950s and 1960s around the automobile and massive per capita consumption of petroleum products.

Of course, all other industrial countries, except Britain, Canada and Norway, have faced the oil crisis. Japan, in particular, is much more dependent on foreign sources of oil and other raw materials than the United States. Of its raw materials 96 percent are imported. To a significant extent, however, Japan and other industrial countries managed to offset the higher price of oil by increasing their manufactured exports, especially to oil-producing countries such as Saudi

Arabia, Venezuela, Nigeria and Indonesia (Magaziner and Reich 1982: 76). The US response was the poorest. In the period 1973–9 the United States covered only 28 percent of its increased oil bill with increased exports. By comparison, West Germany covered 76 percent and France 43 percent.

Higher oil prices did engender some important changes in the US economy. They encouraged conservation, fuel-switching and more petroleum production outside OPEC's control. They also triggered a deep recession from 1979 through to 1982. This led to dramatic drops in oil prices in 1983–4 (*Business Week* 1983: March 7). However, US reserves and production are both declining (*New York Times* 1979: January 7). In the long run, therefore, and as long as oil remains the major fuel of industrialization, the United States will have to depend on foreign sources. This is a major shift for the United States. Unlike the British and the Japanese, whose economies are built on importing raw materials, the United States has always relied almost entirely on domestic sources. After the 1970s this can no longer be the case, especially with oil.

The trends in oil and minerals such as iron ore have been offset to a degree by rapid increases in the US export of agricultural products, especially cereal grains. In the early 1970s the US government undertook the task of stimulating farm (and armament) exports as a way of covering the first of a series of balance of payments deficits beginning in 1971. Even in 1970, before the oil price increases of 1973 and later, the United States was importing 80 percent or more of eight basic raw materials, many food products and large quantities of manufactured goods (Lappé and Collins 1978: 242). Easy financing was arranged for foreign grain sales, and prices were raised temporarily by cutting back on the acreage in production.

The value of food exports from the United States grew from $3,983 million in 1965 to $15,359 million in 1975 and to nearly $30,000 million in 1982. About 65 percent of this value comes from cereal exports. Some government encouragement has come through the use of food aid to 'prime the market.' This induces a reliance on US exports, both physically and through the development of different tastes for different foodstuffs, especially bread in non-wheat-producing countries. The US government has also made large 'grain deals,' especially with the Soviet Union and China.

Since the late 1970s, however, a dramatic increase in food production outside the United States has eroded the position of the United States as a premier grain supplier. American exports plummeted by 15 percent between 1982 and 1983 while shipments by the 15 other largest exporting countries rose by 12 percent. Like the faltering US steel industry, American agriculture is struggling to overcome fierce foreign competition and vast overcapacity – with major implications for the US economy as a whole. For example, government farm out-lays soared 62 percent in 1983 and now constitute a sizeable part of the huge federal deficit; in 1982 farm exports declined by 11 percent and have continued to drop since, worsening the overall US trade deficit; and, finally, the assets of the agricultural sector are also shrink-ing for the first time since the early 1950s, threatening the farmers' ability to service their debts and cutting their demand for consumer and capital goods (*Business Week* 1983: March 21).

American agriculture also faces some troublesome long-range prob-lems. The United States no longer has the excess capacity it once enjoyed in such basic agricultural resources as land, water and energy. Chronic problems such as soil erosion also appear likely to reduce fer-tility in some areas. There is no guarantee that in the future it will be possible to compensate for an ensuing loss of productivity by the genetic improvement of crops, the application of more fertilizer or other technical 'fixes.' These developments, when combined with other factors such as soil compaction, soil salinity and the possibility of climatic change, make American agriculture increasingly vulner-able to disruption (Batie and Healy 1983).

The overall negative position of the United States in raw materials is undeniable. It is attributable in part to the voraciousness of the US economy. But its more recent origins lie in the tempting availability of sources overseas at low cost and high profit. These sources, however, have proven to be more expensive than first thought. The end of the colonial era allowed less developed countries to gain more control over their own resources. Without resorting to its own colonialism the United States must pay the price when these countries collaborate to command a higher price for their products. And, as of 1986, American agricultural exports offer insufficient and unreliable compensation.

The economic drain of Pax Americana

1967 is a key year because of two events. It marked the height of the war in Vietnam and, with hindsight, it marked the beginning of the downturn in American economic growth. These events are not unrelated. The Vietnam War symbolizes the US containment policy in action. The United States was in business to make the world safe for big business. But this was costly. The Vietnam War alone was to cost well over half a trillion dollars (Lappé and Collins 1978: 241).

Throughout the 1950s and 1960s the basic US balance-of-payments deficit was equivalent to the exchange costs of American military forces in Europe (Calleo 1982). In fact, however, it was the Vietnamese who turned the *Pax Americana* into an economic as well as a political disaster. Their failure to roll over and play dead when Americans shot at them turned the war into a financial nightmare. New operations in Vietnam meant new weapons and new troops, both of which cost money. These expenditures, without a corresponding increase in revenues, set off the strongest inflation since 1950 (Mosley 1985).

Although military expenditures declined in the 1970s relative to the peak years of the Vietnam War, America's military costs remained high relative to those of other industrial countries. Much of this was linked to foreign obligations, especially NATO (Alperovitz 1986). But the Nixon, Ford and Carter presidencies were not ones of foreign-policy activism for its own sake as it had been for Kennedy and Johnson in Vietnam. This was not so much foreign retreat, for they all remained committed to *Pax Americana*, as a sense of the new constraints and limitations under which the United States must operate. A spectacular 'loss' for American power in the Carter presidency, the collapse of the Shah's regime in Iran, was to usher in a new era in American national politics: an attempt to go back to the old days when American power and American growth went hand in hand.

Under the Reagan presidency military spending has increased astronomically. Most of the increased spending has gone into so-called 'strategic' weapons such as the MX missile, aircraft carriers, the B-18 bomber and the Trident submarine. Much less has gone into conventional weapons or manpower. This has boosted the US war economy and, at least in the short run, through the vast federal deficits created, stimulated the US economy in general (*Business Week* 1982: November 29; Mintz and Hicks 1984). Between fiscal years 1982 and 1985 military expenditures grew by 32.4 percent in constant dollars. During a comparable period of the Vietnam War (fiscal years 1965 to 1968) the military budget rose 42.7 percent (DeGrasse 1984). So, as in the late 1960s, in the absence of increased revenues to pay for this or of substantial reductions in other forms of federal spending, the long-term consequences will be severe.

Military spending is responsible for a large component of the major contemporary US budget deficit. After subtracting such self-funding programs as social security from the federal budget and adding military-related costs together, 49 percent of the US government's funds were spent on military activities in 1981. In the fiscal year 1986 the US Department of Defense used more than 56 percent of the government's general funds. America's political–military ambitions are now completely outrunning its economic base (DeGrasse 1984).

From one point of view the increases in American military spending are being paid for by foreigners. The huge federal deficit raises interest rates and attracts foreign investment. Current-account deficits must be financed by borrowing from abroad (Schneider 1985). As recently as 1982 US investments abroad exceeded foreign investments in the United States by $169 billion. But by 1985 foreign investments in the United States were $82 billion greater than US foreign investments. That amounts to a shift of $250 billion. In essence the United States has been doing what many Americans have long accused the less developed world of doing: living beyond its means. Consumers are spending and borrowing rather than saving, thus limiting the money available for investment. The US government is vastly overspending relative to what resources it has (Schneider 1985).

The consequences for the US economy are ominous. For years the United States has been able to import more goods than it exports because income flows from foreign investments have offset the trade deficit. But the United States is now paying out more than it is taking in because it is borrowing so heavily overseas. During the nineteenth century the United States was a net debtor country as foreigners invested huge sums in the young nation. Today, however, the United States is borrowing to finance military spending and consumption, *not* private investment. 'Smoke is coming out of the hay pile,' says one economist, 'At some point it will burst into flame. Meanwhile, we're piling on more hay' (*Business Week* 1984: February 27).

The priorities of domestic 'Counter-Keynesianism'

From the late 1940s onwards, American governments preferred to stimulate and stabilize the US economy through indirect fiscal and monetary policies rather than by taking direct action. This is what Wolfe (1981a) has termed 'Counter-Keynesianism' because it reversed the order in the relationship between government and business from government–business to business–government in direct contravention of Keynes' own proposals for government economic management. But American governments shared with Keynes a penchant for demand management – stimulating economic growth through stimulating consumption. One industry was given a high priority: housing and real estate. This happened not because housing was seen as a particularly stimulative industry although it did prove to be, but because, in the battle in the late 1940s over what would remain of the New Deal, legislative action came to focus on housing (Wolfe 1981a).

To supporters of the New Deal a federal housing policy proposing public housing and guaranteed relocation for slum residents removed by urban development was a high priority. But the strength of the opposition was enormous. A compromise eventually resulted, the Housing Act of 1949. This act literally transformed the face of America. As Wolfe (1981a: 84) puts it:

Title I laid out the whole postwar approach to urban housing and became the basis for massive urban renewal programs. Title II redesigned the suburbs, creating insured government mortgages to purchasers of new homes in the Levittowns that were beginning to appear.

In the present context what is important is that the 1949 Act locked the US federal government into a massive stimulus of the US economy through a transformation of the American landscape.

Instead of making a political choice the Housing Act of 1949 bound the US government into a *growth* policy based on providing public subsidies and tax breaks to the private housing industry. It was in Wolfe's (1981a: 88) words, again, 'one of the biggest public works boondoggles in American history.' Throughout the 1950s and 1960s governments continued with policies using the housing and real-estate industry as a 'lever' to reduce the cyclical instability of residential construction and thus reduce cyclical effects within the economy at large. Other policies, such as those relating to the construction of freeways, inadvertently had similar effects. Consequently, a large residential real-estate industry became even larger. One effect of government policies was to expand the size of the population who could afford to buy houses (through mortgage guarantees, insurance, tax breaks, etc.). Another was to create a giant 'sponge' for capital investment (Berkman 1979).

Suburbanization became the engine of growth for the US economy after World War II (Walker 1978). Huge housing subsidies allowed millions of people to acquire suburban houses, creating millions of jobs and large profits for industries involved in construction, earth-moving machinery, building materials, domestic appliances, cars and petroleum products. Massive highway subsidies went to build the rapid-access highways looping and bisecting cities (Checkowoy 1980). Suburbanization also set the basis for the deterioration of central cities. Rising suburban land values were accompanied by declining central-city property values. Throughout the 1950s and

1960s population and employment suburbanized, leaving central cities with declining tax bases and swelling populations of unskilled workers with limited employment opportunities. The suburban stimulus to the American economy coincided with the migration northwards of blacks and poor whites from the rural South displaced by farm mechanization and moved by a desire for a better life. They moved into the central cities. Although there have been signs of central-city revitalization and the movement of some more affluent people back into the central cities, the segregation of poor and rich and black and white between city and suburbia persists to this day (Cox 1973; Browne and Syron 1979).

Some cities are now in severe economic distress. Examples would include Detroit, Cleveland, Cincinnati and St Louis in the North Central region; New Orleans, Louisville, and Birmingham in the South; and Oakland in the West (Browne and Syron 1979). These cities now suffer from both insufficient jobs and serious fiscal problems. Low incomes mean that it is more difficult for residents of these cities to finance a given level of local services (school, police, etc.) than residents of other, more prosperous cities. The 'good life' in the suburbs for some has gone along with bare economic survival for others in declining central cities. How has this happened?

As of 1978 residential fixed investment in the United States averaged only about 5 percent of US GNP, but construction was the largest industry in the United States, real-estate brokering was the nation's largest licensed occupation, and more than half the assets of *all* banking institutions were mortgage or construction loans for 'home building' (Mayer 1978: 6). Despite its rather modest contribution to total output, housing accounts for a large share of credit flows, wealth and indebtedness, creates employment and income in industries such as furniture and domestic appliances, and contributes significantly to cyclical fluctuations in total economic output (Agnew 1981).

Housing is a really attractive investment in the United States. It offers extraordinary tax advantages compared to alternative forms of investment such as corporate stocks, bonds, or direct investment in small businesses. Yet it is these other investments that provide most of the innovation and job creation in the US economy. But a large fraction of the American population has learned to love the deductibility of mortgage payments and property taxes from taxable income. The most incredible tax advantage results from the ability to sell a house without paying capital gains tax on the proceeds as long as another house is purchased and to make $100,000 in capital gains after the age of 55 without paying any tax whatsoever. No other form of investment offers anything remotely approaching these advantages. Furthermore, they encourage people to purchase ever more costly and bigger houses as time goes by.

In the 1970s high rates of inflation multiplied the tax advantages of housing investment. From 1976 until 1980 the price of existing single-family houses increased at an annual average of 12.7 percent. The average house purchased with a 20 percent downpayment thus showed a 63.5 percent annual return on initial investment. Even after deducting borrowing costs, the profits associated with home ownership are incredible – and tax-free.

Consumers are alert to this investment opportunity. Home ownership increased tremendously in the 1970s. But under inflationary conditions this diverted investment from other activities. In the first place the attraction of home ownership discouraged consumers from saving current income. All their savings are in their homes. In the second place one effect of inflation is to push people into higher tax brackets, steadily increasing the value of the tax shelter provided by investment in housing. At the same time, inflation penalizes other types of investment. It harms bonds, for example, because their fixed-interest payments represent a falling rate of return over time.

In the 1970s, therefore, at precisely the time American manufacturing industry needed investment it was being diverted into housing and real estate. This resulted from policy decisions made many years ago when in lieu of a national housing policy the real-estate lobby was let loose in the US Treasury. Domestic growth was made a substitute for political choice. But just as expansion abroad has come back to haunt the United States, so has the domestic growth option based on giving a priority to residential real estate and suburbanization.

Conclusion

There is, then, nothing mysterious about American economic decline in the period 1967–86. A whole set of factors connected in one way or another with the 'coalition' that brought unprecedented growth to the United States in the 1950s and 1960s have with the passage of time put a lid on that growth and precipitated decline.

A 1985 article in the *New York Times Magazine* by Theodore H. White, the late elder statesman of American journalism, missed this point completely. In writing of 'The Danger from Japan,' he charged that the Japanese were 'beating' the United States economically by deceit and nationalistic tricks. He was nostalgic for the time just after World War II when American power was unchallenged.

Commenting on White's article, another American journalist, Anthony Lewis (1985), catches part of the problem in his subtitle 'Looking at Tokyo instead of Ourselves.' He then notes that at least one reason why America now lags behind Japan economically is both deeply ironic and a commentary on where America has come over the past forty years:

While we forced Japan to renounce militarism, we ourselves have become a profoundly militarized society . . . we faced a threat from the Soviet Union. But instead of dealing with it rationally, we have again and again exaggerated the threat, seeing 'missile gaps' when none existed. Now the compulsion to build more and more weapons is fed by the laboratories, the manufacturers, the politicians, the local citizens who fear there is no other source of jobs . . . By all means press Japan to compete fairly. But begin by understanding that America cannot compete effectively while it wastes billions on 'Star Wars' and needless weapons, mortgaging its economic future with immense deficits. The fault, dear Brutus, is not in our stars but in ourselves.

Regional impacts

The relative decline of the American economy since 1967 has had different effects in different regions of the country. This is hardly surprising. The declining relative growth of the American economy has been concentrated in the manufacturing sector. The manufacturing sector is itself geographically concentrated (see Figure 4.10). Seven of the nine states with totals of value-added in manufacturing in excess of $20 billion are within the old manufacturing belt. These are the states of New York, New Jersey, Pennsylvania, Ohio, Michigan, Illinois and Indiana. Two other manufacturing centers are California, with the highest value-added in manufacture of any single state, and Texas, which in 1977 ranked eighth.

The impasse and regional shifts

However, not all industries have been equally affected by the American impasse. In terms of loss of sales within the domestic market, such industries as consumer electronics, textile machinery, footwear, and electrical components have been especially hard hit (see Table 4.8). But even the relatively smaller declines in such industries as cars and clothing are important because of the absolute size of these industries. They are major employers and sources of value-added in

Table 4.8. *Industries most affected in the domestic market by foreign penetration, 1960–79*

Domestic market	Ranked by total sales of industry (% of market)		
	1960	1970	1979
Autos	95.9	82.8	79.0
Steel	95.8	87.5	86.0
Apparel	98.2	94.8	90.0
Electrical components	99.5	94.4	79.9
Farm machinery	92.3	92.2	84.7
Industrial inorganic chemicals	98.0	91.5	81.0
Consumer electronics	94.4	68.4	49.4
Footwear	97.7	85.4	62.7
Metal-cutting machine tools	96.7	89.4	73.6
Food-processing machinery	97.0	91.9	81.3
Metal-forming machine tools	96.8	93.2	75.4
Textile machinery	93.4	67.1	54.5
Calculating and adding machines	95.0	63.8	56.9

Source: Business Week 1980: June 30.

manufacture. The picture in export markets is similar if perhaps even more bleak (see Table 4.9). Many US export markets shrank from 1962 to 1979 by more than one-half. The declines are especially severe and important from an employment perspective in motor vehicles, chemicals and agricultural machinery. Altogether, American industries lost 23 percent of their share of world exports in the 1970s. These losses are all the more important in that they came about at a time when the US dollar depreciated in value by about 40 percent. This should have made American exports cheaper and foreign imports more expensive.

In both absolute and relative terms manufactured exports are most important to the states of the manufacturing belt, especially Massachusetts, Connecticut, Pennsylvania, Ohio, Michigan, Illinois, Indiana and Wisconsin (see Figure 4.11). But California, with only 37 percent of its employment related to manufactured exports, still has the largest number of jobs in this category of any single state (320,000). Employment in the United States relating to the export of manufactured goods accounted for 3.5 million jobs in 1976 or 11 percent of total manufacturing employment. The upshot of all this is that the impact of declining economic growth has been most directly felt in the 'traditional' manufacturing areas of the Northeast, the Midwest, and California.

Table 4.9. *Industries most affected in the export market by foreign penetration, 1962–79*

Export market	Ranked by size of US exports (% of world exports)		
	1962	1970	1979
Motor vehicles	22.6	17.5	13.9
Aircraft	70.9	66.5	58.0
Organic chemicals	20.5	25.7	15.0
Telecommunications apparatus	28.5	15.2	14.5
Plastic materials	27.8	17.3	13.0
Machinery and appliances (non-electric)	27.9	24.1	19.6
Medical and pharmaceutical products	27.6	17.5	16.9
Metal-working machinery	32.5	16.8	21.7
Agricultural machinery	40.2	29.6	23.3
Hand or machine tools	20.5	19.1	14.0
Textile and leather machinery	15.5	9.9	6.6
Railways vehicles	34.8	18.4	11.6
Housing fixtures	22.8	12.0	8.1

Source: Business Week 1980: June 30.

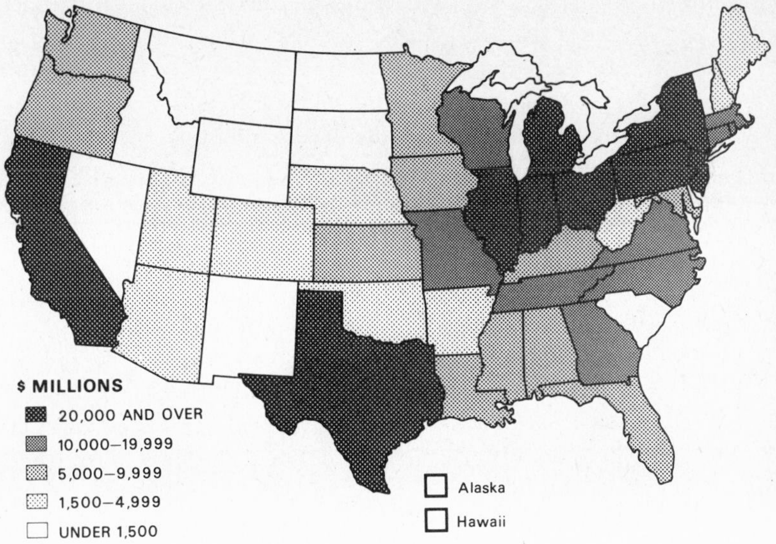

$ MILLIONS
- 20,000 AND OVER
- 10,000–19,999
- 5,000–9,999
- 1,500–4,999
- UNDER 1,500

Alaska
Hawaii

Figure 4.10: Value-added in manufacturing by state, 1977 (*Business America*, November 19, 1979)

Table 4.10. *Manufacturing jobs in the manufacturing belt, 1967–80*

	1967 (000s)	1980 (000s)	% change
Middle Atlantic			
New York	1,929	1,451	−24.8
New Jersey	881	783	−11.1
Pennsylvania	1,550	1,328	−14.3
East North Central			
Ohio	1,397	1,268	−9.2
Indiana	710	658	−7.3
Illinois	1,397	1,222	−12.5
Michigan	1,134	1,007	−11.2
Wisconsin	512	560	+9.4
New England			
Maine	111	113	+1.8
New Hampshire	95	117	+23.2
Vermont	43	51	+18.6
Massachusetts	714	673	−5.7
Rhode Island	122	128	+4.9
Connecticut	478	442	−7.5

Source: ACIR 1980.

One major consequence has been a loss of manufacturing jobs. In the manufacturing belt losses have been especially severe in New York, Pennsylvania, Illinois and Michigan (see Table 4.10). Some states have fared somewhat better but these, especially those in New England, are states with much smaller absolute levels of employment in manufacturing. It is important to put the losses of employment in context, however. New York, for example, is still the second-ranked manufacturing state after California, accounting in 1979 for 7.6 percent of total national value-added by manufacturing.

Another significant consequence has been the expansion of manufacturing employment outside the regions with which it has been hitherto synonymous. Indeed, there are today as many jobs in manufacturing outside the manufacturing belt as inside it (see Figure 4.12). This reflects a trend that was under way before the 1970s but nevertheless quickened during that decade. By way of example, Texas has acquired much new manufacturing investment during the period of national downturn. In the period 1967–80 Texas registered the greatest absolute increase in manufacturing employment of any state. The three largest investing industries have been chemicals and allied products, petroleum products, and primary metals. Most importantly, much of the growth has been in manufactures not strongly tied

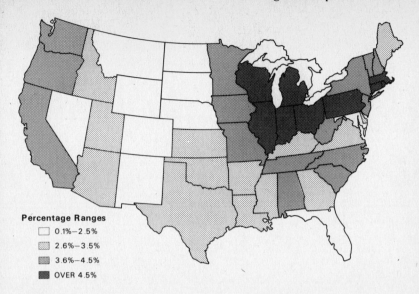

Figure 4.11: Employment related to manufacturing exports as a percentage of total civilian employment by state, 1976 (*Business America*, November 19, 1979)

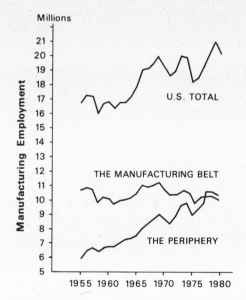

Figure 4.12: The geographical shift in manufacturing employment, 1955–80 (ACIR 1980)

Table 4.11. *Jobs created and lost through openings, closings, relocations, expansions and contractions of private business establishments in the United States, 1969–76 (thousands of jobs)*

	Number of jobs in 1969	Jobs created		Jobs destroyed		Net job change[d]
		By openings and immigrations[a]	Expansions[b]	By closures and outmigrations[c]	Contractions[b]	
US as a whole	57,936.1	25,281.3	19,056.1	22,302.3	13,183.2	8,851.9
Frostbelt	32,701.2	11,321.5	9,470.4	11,351.7	7,212.1	2,228.1
Northeast	15,824.6	4,940.4	4,347.5	5,881.5	3,589.0	−182.6
New England	3,905.3	1,251.2	1,131.0	1,437.2	952.1	− 7.1
Middle Atlantic	11,919.3	3,689.2	3,216.5	4,444.3	2,636.9	−175.5
Midwest	1,686.6	6,381.1	5,123.0	5,470.2	3,623.2	2,410.7
East North Central	12,563.6	4,670.6	3,581.8	3,962.6	2,651.7	1,638.1
West North Central	4,313.0	1,710.6	1,541.2	1,507.6	971.5	772.7
Sunbelt	25,234.9	13,959.8	9,585.7	10,950.5	5,971.0	6,624.0
South	16,044.5	8,934.2	5,964.6	6,824.3	3,803.3	4,271.2
South Atlantic	8,204.1	4,651.2	2,913.0	3,547.9	2,014.2	2,002.1
East South Central	3,065.2	1,518.2	1,089.9	1,211.0	631.9	765.2
West South Central	4,775.2	2,764.8	1,961.7	2,065.4	1,157.2	1,503.9
West	9,190.4	5,026.6	3,621.1	4,126.2	2,167.8	2,352.7
Mountain	1,941.9	1,226.1	953.6	977.9	481.0	720.8
Pacific	7,248.5	3,799.6	2,667.6	3,148.3	1,686.8	1,632.1

to energy resources. 55 percent of manufacturing employment is now in durable-goods production in contrast to 45 percent in 1960. Texas has also experienced growth in its steel industry. There are a number of so-called 'minimills' in Texas producing steel from scrap in electric furnaces. At first they focused on the Texas market but now they are expanding both in areas and products (*Business Week* 1983: June 13).

The relative importance of manufacturing industry to Texas and other states experiencing a boom in manufacturing investment must be kept in perspective. In 1980 manufacturing employment was only 18 percent of total non-agricultural employment in Texas compared to a national figure of 22 percent. What is most remarkable is the rate of increase given the dismal picture in the United States as a whole.

One way of examining overall regional trends in economic growth is to look at patterns of job 'openings' and 'closings.' Bluestone and Harrison (1982) report that between 1969 and 1976 private investment in new plants and offices created 25 million jobs. At the same time, shutdowns had removed 39 percent of the jobs that had existed in 1969 – around 22 million jobs. The relationship between the openings and closings varied significantly between regions (see Table 4.11). New jobs exceeded job losses in the Midwest and in the so-called 'Sunbelt' states. But in the Northeast and in some Sunbelt states (e.g. Idaho and Utah) more jobs disappeared than were created. In the late 1970s the Midwest or East North Central region joined them (Cloos and Cummins 1984). Especially surprising is the situation in the South. The overall pace of economic growth in this region has been much higher than elsewhere. Even so, between 1969 and 1976, 7 million jobs were lost as a direct result of shutdowns and another 3.8 million disappeared through cutbacks in existing operations. New jobs more than compensated for this, but the *volatility* is notable.

Notes to Table 4.11

[a] Bluestone and Harrison aggregate openings (or start-ups, or what Birch calls 'births') and 'immigrations' (plants that are new to the area but that are known to have previously existed elsewhere) into a single category.
[b] These columns refer to employment change in establishments that neither relocate nor shut down during the period of analysis.
[c] They aggregate 'closures' (or shutdowns, or what Birch calls 'deaths') and 'outmigrations' (plants that previously operated in the area, closed there, and then reported elsewhere) into a single category.
[d] Because the employment change associated with recorded relations (immigrations and outmigrations) is so small – between 0.2 percent and 2.0 percent of net employed change over any particular period of time in any state – there is some, but probably little, double-counting in these regional and national totals.
Source: Bluestone and Harrison 1982: 30 (based on Birch 1979).

Table 4.12. *Closings in a sample of manufacturing plants with more than 100 employees that were open on December 31, 1969 but closed by December 31, 1976*

	Number of states	Percentage of US population	Number of plants in the sample in 1969	Number in the sample closed by 1976	Proportion of 1969 plants that closed by 1976	Interregional percentage distribution of closings
Northeast	9	24.1	4,576	1,437	0.31	38.6
North Central	12	27.8	3,617	904	0.25	24.2
South	16	31.0	3,101	1,042	0.34	28.0
West	13	17.1	1,155	344	0.30	9.2
Total	50	100.0	12,449	3,727	0.30	100.0

Source: Bluestone and Harrison 1982: 32.

Table 4.13. *Growth in regional stocks of manufacturing capital,
1960–76 (annual % rates of change)*

	Northeast	North Central	South	West	US
1960–5	2.0	2.3	3.7	3.7	2.7
1965–70	4.4	4.5	6.8	5.6	5.2
1970–6	2.0	2.5	5.3	4.0	3.4
1960–76	2.7	3.0	5.3	4.4	3.8

Source: Browne *et al.* 1980: 7.

If attention is directed towards larger manufacturing plants, the
Northeast appears as the major loser (see Table 4.12). With 24 percent
of the country's population in 1970, the Northeast experienced
39 percent of the shutdowns of the largest manufacturing facilities.
But the South also suffered a dramatic loss of large plants during the
same period.

Much of the change in employment patterns, however, is at a more
local scale. For example, during the late 1960s and early 1970s, Massachusetts was considered a case of terminal industrial decline. Yet
just over its border, in New Hampshire, various towns were
experiencing unprecedented booms in industrial employment. Like-
wise in the South. While Houston, Dallas and Tulsa were overflowing
with new employment in oil, chemicals and manufacturing industry,
some areas, such as the Piedmont region of South Carolina, were
declining. After World War II the textile industry of the South was
relocating overseas (Schmidt 1986).

On the whole, however, the South is the region of the United States
in which the rate of capital investment has been consistently highest
since the 1960s (see Table 4.13). In the 1960s rates of capital accumulation in the different regions were similar. In the 1970s the growth in
capital slowed but the slowdown varied considerably across regions.
The South has been lightly affected, while the rate of capital expansion
in the Northeast in the period 1970–6 was cut in half.

Within the four large regions, rates of capital investment vary considerably between states (see Figure 4.13). In the Northeast the New
England states have generally experienced more growth in investment
than the three Middle Atlantic states. The northern New England
states have fared better than the southern ones. Some southern states
– Maryland, Delaware and West Virginia, for example – experienced
lower rates of investment than states such as Florida, Texas, Alabama

Table 4.14. *Regional migration patterns, 1970–4 and 1975–9 (in thousands)*

	Northeast	North Central	South	West
1970–4:				
Immigration	1,035	1,800	3,377	2,141
Outmigration	1,993	2,512	2,312	1,536
Net migration	−958	−712	+1,065	+605
1975–9:				
Immigration	1,035	1,830	3,583	2,552
Outmigration	2,138	2,737	2,513	1,615
Net migration	−1,103	−907	+1,070	+937

Source: Bluestone and Harrison 1982: 99.

and Georgia. In the West, some of the mountain states – for example, Wyoming and Montana – received very little manufacturing investment, while Arizona and Colorado boomed. However, despite these variations and at least through until the mid-1970s, capital investment in the South has surpassed that in all but a few states elsewhere, especially the states of the Northeast and East North Central – the old manufacturing belt (Browne *et al.* 1980).

The regional pattern of capital investment has had important implications for population redistribution within the United States. For those living in places experiencing plant closings one alternative is to move to places where jobs are available in greater quantity. Such movement has been on a vast scale. Until 1970 the migration of people from the Northeast and North Central regions to other regions was generally small. In the 1970s, however, the migration accelerated rapidly to a rate of at least 400,000 per year (Vining 1982). Between 1970 and 1979 almost 7 million people moved to the South. Another 4 million moved to the West. From population movements alone, including migration from the South and West to the other regions, there was a *net* gain of 3.7 million for the South and West in the 1970s (see Table 4.14).

The tide of migrants from the old industrial core to the South and West has been supplemented by a flood of immigrants from Latin America, especially Mexico. Many of these are illegal or 'undocumented' immigrants. The Census Bureau estimates that there were, in 1984, 5 million to 7.5 million illegal residents in the United States holding between 4 million and 6 million jobs (*Business Week* 1984: May 14). These people are extremely popular with labor-intensive industries which make virtue, and profits, out of their perilous situation. The pattern of illegal or undocumented Mexicans is a highly concentrated one. More than 50 percent of the immigrants are in California. However, there is evidence of recent dispersal into other regions (Jones 1984).

Overall, the interregional flows of migrants represent a summary measure of the more profound shifts of capital that have been taking place within the US economy (Browne 1979). The South benefited in the 1960s and 1970s from rapid investment. In contrast, investment in the Northeast lagged behind. This pattern was shared, to a certain extent, by other traditionally high-income areas – the East North Central and Pacific states. The results were, in the South, a relative increase in the proportion of the population employed, attributable to lower unemployment rates and higher labor force participation rates, and an increase in relative earnings (Browne *et al.* 1980). In the Northeast as a whole the proportion employed declined relative to the national average and, in New England, relative earnings fell (Browne *et al.* 1980). Two regions were especially sensitive to the relative decline of the United States within the world-economy. The East North Central region specializes in durable-goods manufacturing and

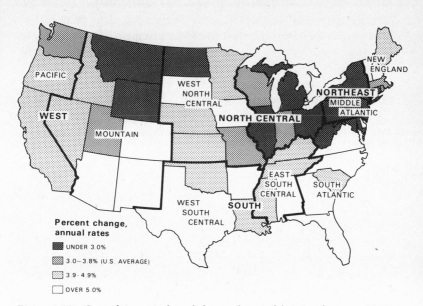

Figure 4.13: Growth in capital stock for total manufacturing, by state, 1960–76 (Browne *et al.* 1980: 8). Reproduced from the *New England Economic Review* (Federal Reserve Bank of Boston) with permission

the Northeast in traditional 'smokestack' or heavy industries. It is these two regions – the manufacturing belt – that have declined the most: they have lost people, have declining rates of capital investment, have had higher unemployment rates and have lower relative earnings.

The appeal of the periphery

All sorts of factors have been invoked to explain the relative growth of the southern and western peripheries compared to the manufacturing belt. To name just a few in no particular order: the much greater social welfare burdens carried by the North; federal policies that discriminate against the North because of the historical southern domination of the powerful congressional committee system; the widespread use of air-conditioning in the South, which made the climate there more congenial to northerners and productivity higher among the indigenous population; the depletion of the local resource base in the manufacturing belt; and the secular decline in transportation costs that has allowed industrial production to move away from primary markets into areas with cost advantages (McManus 1976).

If the previous discussion in this chapter has been to the point, however, the primary cause of manufacturing investment and employment shifting from core to peripheries lies in the adjustment strategies of firms rather than federal expenditures or the diffusion of air-conditioning. Manufacturing job decentralization seems to have led the process of Sunbelt growth. This is *not* to say that federal government policies and expenditures were unimportant. Far from it. They have *enabled* the process of relocation to take place. But in and of themselves, without the stimulus of the crisis in the American economy as a whole, they would not have been sufficient to account for the dramatic regional shifts noted in the last section.

But what did locations in the periphery have to offer? In the first place, and especially in parts of the South, there were lower labor costs. One in three manufacturing jobs in the South in 1978 was in an industry in which hourly earnings were below the national average for all production jobs (Tabb 1984: 9). In 1978 half of all manufacturing employees nationwide were in industries with average wages above $6.50 an hour, but only one-third of southern workers held such jobs (Rones 1980: 14). Many of these workers are employed in so-called 'high-tech' industries producing such items as semiconductors, calculators and computers. High technology does not mean high pay, often the opposite.

Table 4.15. *Average prices of residential energy sources delivered to users, nationwide and by region, April 1978 to March 1979 (dollars per million Btus)*

	Electricity	Natural gas	Fuel oil and kerosene	Liquefied petroleum gas
Nationwide	12.10	2.74	3.93	5.09
Northeast	15.34	3.42	3.98	7.93
North Central	13.64	2.57	3.82	4.55
South	11.75	2.85	3.94	5.16
West	8.28	2.30	3.77	4.18

Source: US Department of Energy 1980: Table 3.

But not only low-wage industries have relocated or, in quantity more importantly, set up from scratch in the South. In fact it has been in fast-growing, high-wage industries that the South's growth has been most spectacular. However, in the South they need pay *relatively* less than they would have to in the Northeast. This is clearly the case in Texas, Kentucky and Florida, three states with below-average wages but a balance of high-wage industries (Browne 1984). Why? Because these are so-called right-to-work states, like many others in the Sunbelt, in which closed shops, operated under a contractual agreement between a union and an employer requiring the employment of union members only, are illegal. In 1982 Texas ranked forty-seventh among the states in the proportion of its workforce that is unionized – a scant 13 percent. The lack of a pro-union environment can mean lower wages and a generally more favorable 'climate' for business than that in most northeastern states (Weinstein and Firestine 1978; Newman 1984).

The 1970s was a decade in which energy prices came to figure alongside labor costs as a major item in a corporate locational calculus. Prior to 1973 energy prices were generally low throughout the United States. But after that point and until 1985, major regional disparities appeared (see Table 4.15). In terms of residential energy prices consumers in the Northeast and North Central regions faced considerably higher prices than did those in the South and West. This is most pronounced for electricity, for which the Northeast pays 85 percent higher rates than the West (where cheap hydro power keeps down the average). For natural gas the Northeast pays nearly 50 percent more. However, *rates of increase* in energy costs have not been that different across regions (see Table 4.16). Some regions, such as the South, now

Table 4.16. *Average energy prices by region,
1970 and 1980 (dollars per billion Btus)*

	1970	1980	% increase 1970–80
Residential			
Northeast	1,598	5,808	263
Midwest	1,430	4,388	207
South	1,411	4,136	193
West	1,098	3,603	228
United States	1,403	4,472	219
Industrial			
Northeast	847	4,256	402
Midwest	723	3,130	333
South	462	2,795	505
West	651	3,167	386
United States	628	3,166	403

Source: Northeast–Midwest Institute 1981.

have less relative advantage in industrial energy costs over other regions than they once had. Though of course climatic differences lead to higher absolute energy costs in the Northeast, the spread of air-conditioning in the South exerts its influence on energy costs, too (Ellis 1979). In addition, regional differences in industry mix and factor substitution patterns suggest that only certain manufacturing sectors are especially dependent on energy supply in general and specific fuels in particular (Lakshmanan *et al.* 1984). This should be borne in mind in the following discussion.

Bensel (1984: 264–6) has combined the union shop/right-to-work dichotomy with the geography of relative energy disadvantage. He demonstrates that after 1972 there is a marked divergence in population growth – a surrogate for economic growth – between the ten most energy-disadvantaged (high cost and large imports), union shop states and the ten most energy-rich, right-to-work states (see Figure 4.14). The combination of an anti-union environment and energy-cost advantages offers a most plausible explanation for the increasing appeal of the periphery in the 1970s (see also Browne *et al.* 1980).

But undoubtedly other factors have been at work, too. Most research on US regional economic growth has found little evidence that incentives offered by state and local governments, such as tax credits, loan guarantees, industrial development bonds and low interest loans, affect the locational decisions of firms (Weinstein *et al.* 1985). However, inter-state competition in business tax rates has probably been of some importance in creating 'business climates' favorable to industrial expansion (Newman 1984). The overall level and structure of state and local taxes have been important influences on the economic performance of a given area. A number of studies point to this conclusion (Genetski and Chin 1978; Newman 1984). By the mid 1970s, however, the actual differences between states in different regions in levels of business taxation were much reduced as a direct result of inter-state competition (see Table 4.17). Other elements critical to the definition of business climate are of continuing importance (Wasylenko and McGuire 1985). Among these are personal tax rates (Ecker and Syron 1979) and laws having to do with minimum wages, affirmative action and income maintenance

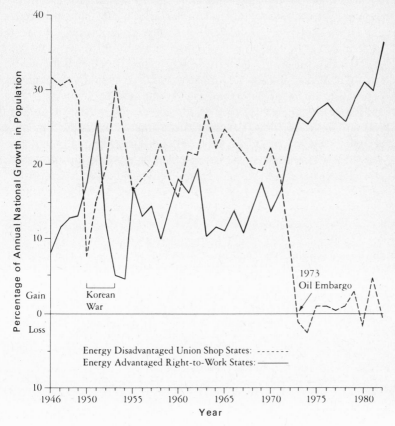

Figure 4.14: Union shop status/energy advantage and percentage change in annual national growth in population, 1946–82 (Bensel 1984: 265). Reproduced by permission of University of Wisconsin Press

(unemployment insurance, food stamps and welfare) (Bluestone and Harrison 1982). The northern industrial states tend to have higher personal tax rates and much greater provision of public goods and services. They therefore provide a less favorable climate for capital accumulation, and thus business activity, than do the 'growing' states of the South and the West. The periphery provides businesses with at least a partial escape from the New Deal without their having to move abroad.

Federal expenditures and tax policies have generally provided a favorable framework for the regional restructuring of the US economy in the 1970s and 1980s. In the early 1970s certain states in the periphery were particularly favored by federal expenditures, especially in those categories most likely to encourage future economic growth (Fainstein and Fainstein 1976). If California, Texas and Florida are compared to New York, Pennsylvania and New Jersey in terms of those expenditures which have the greatest multiplier effects – defense spending, farm subsidies, highway construction and federal civilian employment – the peripheral states had a great advantage. Only in the areas of housing, urban development and public assistance did the three 'core' states have a clear edge. However, regional patterns can be overdramatized when samples are limited to relatively few states. In the area of defense spending, for example, although states such as California, Texas and Florida have been favored, so have certain core states (see Table 4.18). States such as Connecticut and Massachusetts have received as much per capita in defense contracts as have the most favored peripheral states (see also Rees and Weinstein 1983).

Perhaps the best way to estimate the total effects of federal expenditures is to examine the extent to which expenditures exceed the federal revenues raised in a specific state (see Table 4.19). This measures the net dollar flow from the federal government to the various states. The big losers from this perspective are the Great Lakes states (Ohio, Indiana, Illinois, Michigan and Wisconsin) and, to a lesser extent, the Middle Atlantic states. The big winners are the Pacific states (especially California and Washington) and the South Central region (Rees and Weinstein 1983; Erdevig 1984).

In addition to federal expenditures, the federal tax structure has also encouraged regional restructuring. As it stands, the present federal income tax code allows companies to write off old plants, counts the cost of relocation as a business expense, and offers generous tax write-offs on plant and equipment. These are all recent innovations, most dating from 1962 (Luger 1984). Since federal tax

Table 4.17. *State and local effective tax rates[a] by state, 1975[b] (%)*

Region and state	Effective tax rate	Region and state	Effective tax rate
New England	2.46	East South Central	0.96
Maine	2.22	Kentucky	1.21
New Hampshire	2.45	Tennessee	0.94
Vermont	2.77	Alabama	0.63
Massachusetts	2.75	Mississippi	1.16
Rhode Island	1.98	West South Central	1.24
Connecticut	2.09	Arkansas	1.14
Middle Atlantic	1.99	Louisiana	0.96
New York	2.16	Oklahoma	1.34
New Jersey	1.99	Texas	1.31
Pennsylvania	1.66	Mountain	1.91
East North Central	1.54	Montana	3.41
Ohio	1.31	Idaho	1.92
Indiana	1.52	Wyoming	3.51
Illinois	1.50	Colorado	1.63
Michigan	1.71	New Mexico	1.64
Wisconsin	1.89	Arizona	2.01
West North Central	1.74	Utah	1.43
Minnesota	1.83	Nevada	1.46
Iowa	1.99	Pacific	2.22
Missouri	1.12	Washington	1.24
North Dakota	2.47	Oregon	1.69
South Dakota	2.96	California	2.43
Nebraska	1.99	Alaska	3.72
Kansas	2.20	Hawaii	1.72
South Atlantic	1.21	Average	1.69
Delaware	0.85		
Maryland	1.46		
Washington, DC	1.44		
Virginia	1.37		
West Virginia	1.16		
North Carolina	0.96		
South Carolina	1.09		
Georgia	1.10		
Florida	1.34		

[a] Tax rate = (Corporate income tax revenue + business property taxes)

(Value of manufacturing shipments + total retail sales + total wholesale sales + total selected service industry receipts)

This 'business tax rate' excludes unemployment insurance, workers' compensation, and special excise taxes imposed by state or local governments.

[b] Because of data limitations, the tax rates do not strictly correspond to a given year. Corporate income tax data are for 1977; business property tax revenue information corresponds to 1975; total value of sales/shipments/receipts is for 1972. Since the last of these values (at least in current dollars) has increased since 1972, the tax rates as calculated here overestimate actual 1975 tax rates. This should not seriously affect the differences between states.

Sources: Bluestone and Harrison 1982: 186. Tax rate estimates are based on data in Vaughan (1979: Table 14, pp. 74–5) and data on manufacturing, retail trade, wholesale trade and selected services in US Bureau of Census (1975: Tables 11, 1267, 1316, 1323, 1331).

incentives add to recipient firms' after-tax cash flow, they directly encourage disinvestment or plant closure and relocation or expansion elsewhere as business strategies for increasing the rate of profit. As Vaughan (1977: 38) puts it: 'tax credits, applied without regional targeting, generally represent a subsidy to growth areas . . . at the expense of those areas that are growing more slowly.'

By way of summary, the appeal of the American periphery to business in the period 1967–85 has not been dissimilar to the appeal of locations abroad. Above all, the relatively lower-wage, anti-union, lower-public-spending, energy-advantaged local environments in the periphery have provided a suitable alternative to locations overseas for new firms and for many industries 'on the run' from the old American industrial core. Other factors, especially federal spending and federal tax policies have facilitated, if not encouraged, the regional restructuring of the US economy.

The decline of the core?

In 1975 President Ford told New York City, as the *New York Daily News* put it in a famous headline, to 'drop dead.' By opposing federal aid to the near-bankrupt city, Ford was voicing a widespread sentiment among politicians and journalists. The entire Northeast, especially the states from Maine to New Jersey that were the original hub of American commerce and industry, was viewed as an old and stagnant region with little hope of revitalization. In contrast, the Sunbelt, especially the states of the Southwest plus Texas and Florida, was the glamorous growth region with an unlimited potential. In 1986 the picture seems considerably more complex.

In the first place, only parts of the core now seem to be in long-term decline. The East North Central or Great Lakes region, with its very high proportion of durable-goods manufacturing, has been most exposed to changes in the world-economy without the compensation of high levels of defense or other federal spending (Syron 1978; Browne 1978; Cloos and Cummins 1984; Browne 1983). Other parts of the industrial core, especially New England, have rebounded in the late 1970s and early 1980s relative to both earlier performance and to the rest of the country (*Business Week* 1984: January 23). New England in particular has benefited from a number of recent trends. One of these is the increased defense spending of the period 1978–85 (Browne and Hekman 1981). Connecticut and Massachusetts are among the top five states in defense contracts. Another is relatively high levels of foreign direct investment (Little 1985). Foreign-owned

Table 4.18. *Per capita federal spending by program for states ($),*
1975

	Defense contracts	Defense salaries	Highways and sewers	Welfare programs	Retirement programs
Northeast	226	58	39	137	403
New England	382	70	48	118	402
Maine	54	87	59	149	440
New Hampshire	244	195	91	79	435
Vermont	262	35	60	142	395
Massachusetts	314	56	46	128	403
Rhode Island	90	158	40	143	464
Connecticut	763	36	32	86	361
Middle Atlantic	175	55	36	143	403
New York	312	31	27	181	394
New Jersey	146	77	39	105	387
Pennsylvania	135	77	47	108	427
Midwest	120	63	56	96	371
Great Lakes	96	52	54	101	359
Great Plains	177	91	60	83	403
South	164	194	51	113	396
South Atlantic	161	228	57	102	426
South Central	168	161	46	124	368
West	329	210	67	119	394
Mountain	174	214	95	82	385
Pacific	385	209	57	132	398
Total United States	201	132	54	115	392

Source: McManus 1976: 350.

businesses account for a much higher share of jobs and plants in New
England than in most other regions and the United States as a whole.
Perhaps most importantly, wage rates have been below the national
average (whereas in the early 1970s they were 7 to 10 percent above)
and competitive relative to America's 'growth regions' (see Table
4.20). From 1970 until 1975 earnings, responding to the pressure of
high unemployment (10.2 percent in 1975), grew much more slowly
than in the country as a whole (Browne and Hekman 1981; Browne
1984).

But a general trend shared by many of the states of the Northeast
has been the shift towards so-called 'high-technology' industries.
These are industries with a high proportion of technical workers relative to others and a high rate of research and development spending
relative to company sales (Weinstein *et al.* 1985). Research on the
geographical incidence of R & D activities suggests that the Northeast

Table 4.19. *Winners and losers in the flow of federal dollars, 1975*

	Spending per person	Taxes per person	Spending–taxes ratio	Dollar flow (in millions)
Northeast	1,361	1,579	0.86	−10,776
New England	1,470	1,533	0.96	−762
Maine	1,206	1,075	1.12	139
New Hampshire	1,399	1,399	1.00	1
Vermont	1,360	1,167	1.17	91
Massachusetts	1,456	1,535	0.95	−462
Rhode Island	1,342	1,457	0.92	−107
Connecticut	1,663	1,800	0.92	−425
Middle Atlantic	1,325	1,594	0.83	−10,013
New York	1,449	1,636	0.89	−3,392
New Jersey	1,154	1,760	0.66	−4,436
Pennsylvania	1,241	1,426	0.87	−2,185
Midwest	1,128	1,477	0.76	−20,074
Great Lakes	1,064	1,518	0.70	−18,618
Ohio	1,010	1,441	0.70	−4,634
Indiana	1,027	1,411	0.73	−2,036
Illinois	1,230	1,704	0.72	−5,290
Michigan	996	1,539	0.65	−4,971
Wisconsin	996	1,331	0.73	−1,686
Great Plains	1,287	1,374	0.94	−1,456
Minnesota	1,144	1,382	0.83	−934
Iowa	970	1,405	0.69	−1,249
Missouri	1,500	1,362	1.10	657
Kansas	1,398	1,432	0.98	−78
Nebraska	1,193	1,420	0.84	−351
South Dakota	1,395	1,081	1.29	251
North Dakota	1,734	1,288	1.35	283
South	1,389	1,219	1.14	11,522
South Atlantic	1,454	1,303	1.12	4,986
Delaware	1,145	1,743	0.66	−347
Maryland	1,933	1,615	1.20	1,299
Virginia	1,809	1,355	1.34	2,257
West Virginia	1,318	1,091	1.21	410
North Carolina	1,124	1,145	0.98	−115
South Carolina	1,240	1,041	1.19	561
Georgia	1,403	1,217	1.16	912
Florida	1,379	1,378	1.00	9
South Central	1,327	1,137	1.17	6,536
Kentucky	1,327	1,094	1.21	790
Tennessee	1,296	1,147	1.13	627
Alabama	1,374	1,026	1.34	1,255
Mississippi	1.599	908	1.76	1,621
Louisiana	1,236	1,064	1.16	652
Arkansas	1,202	970	1.37	492
Oklahoma	1,443	1,181	1.22	711
Texas	1,296	1,264	1.03	388

Table 4.19 (cont.)

	Spending per person	Taxes per person	Spending–taxes ratio	Dollar flow (in millions)
West	1,172	1,431	1.20	10,639
Mountain	1,651	1,238	1.30	3,631
Montana	1,512	1,183	1.28	246
Idaho	1,358	1,087	1.25	223
Wyoming	1,569	1,295	1.21	102
Colorado	1,646	1,368	1.20	704
Utah	1,449	1,072	1.35	455
Nevada	1,544	1,612	0.96	−40
Arizona	1,639	1,256	1.41	853
New Mexico	1,974	1,024	1.93	1,090
Pacific	1,745	1,497	1.17	7,008
California	1,700	1,526	1.11	3,684
Oregon	1,282	1,371	0.94	−202
Washington	1,968	1,403	1.40	2,008
Alaska	3,736	1,530	2.60	776
Hawaii	2,347	1,490	1.58	741
DC	13,957	1,820	7.67	8,690
Total United States	1,412	1,412	1.00	0

Source: McManus 1976: 349.

has maintained its predominance as the region of greatest R & D concentration (Malecki 1980). Some other research on the distribution of high-technology jobs suggests that this distribution corresponds closely to that of employment in all industries, that is high-technology industries are spread across the country in rough proportion to the distribution of all industries (Armington *et al.* 1983).The Northeast in fact has a higher concentration of high-technology employment than of industrial employment in general (see Table 4.21). Since Armington's study also related 'branch plants' to headquarters, it was possible to show that high-technology plants in the South were owned by out-of-region companies to a much higher degree than such plants in other regions, especially the Northeast.

On the 'down side,' however, the rate of formation of new high-technology firms has been lower recently in the Northeast than in the South and the West (Armington *et al.* 1983). The long-term picture, therefore, is not so bright. But the fact that many of the 'old' industrial states have so many high-tech jobs shows that they cannot as yet be written off as being at the end of their 'life-cycle' (see Table 4.22).

One case study in resurgence is New York City itself. From 1969 to 1976 employment dropped sharply. From 1977 through to 1983,

Table 4.20. *Average hourly earnings of manufacturing production workers in selected states relative to US average (%), 1973–82*

	1973	1975	1977	1979	1982
New England					
Connecticut	101.2	99.0	97.9	96.0	96.8
Massachusetts	95.1	92.8	90.3	89.3	89.3
New Hampshire	82.9	82.2	80.3	80.1	81.8
Rhode Island	82.4	79.5	77.3	76.1	77.8
Middle Atlantic					
New Jersey	104.2	103.3	102.5	100.2	102.3
New York	102.7	101.7	99.8	98.1	98.2
Pennsylvania	101.7	103.1	103.0	104.0	101.5
East North Central					
Illinois	111.7	114.5	110.6	109.0	109.5
Indiana	113.7	113.7	116.2	116.3	115.2
Michigan	na	na	132.7	130.3	131.5
Ohio	116.4	115.3	118.7	117.0	118.5
Wisconsin	108.8	108.9	108.5	108.5	110.2
West North Central					
Iowa	na	na	113.2	115.7	117.6
Minnesota	103.2	105.6	105.1	103.4	107.2
Missouri	99.0	99.4	101.2	100.0	99.5
South Atlantic					
Florida	84.4	85.1	81.5	81.8	82.6
Georgia	79.5	87.7	78.5	79.1	79.4
Maryland	103.2	104.3	106.5	105.8	103.3
North Carolina	73.1	72.8	72.2	72.7	74.7
South Carolina	74.1	74.3	75.4	76.1	78.6
Virginia	81.7	82.6	82.6	83.3	86.6
East South Central					
Alabama	83.6	84.9	86.1	88.8	86.2
Kentucky	97.8	98.8	100.2	101.0	98.6
Mississippi	72.1	74.1	73.1	73.9	75.4
Tennessee	80.4	81.4	82.4	83.0	84.2
West South Central					
Arkansas	73.1	76.4	75.7	77.5	87.7
Texas	na	na	95.4	96.4	101.2
Pacific					
California	108.6	108.1	105.6	104.9	108.7
Standard deviation	13.7	14.1	16.0	15.4	13.9

Source: Browne 1984: 42.

Table 4.21. *High-technology employment growth shares by region,*
1976–1980 (%)

	High technology		All industries	
	Share of employment	Share of growth	Share of employment	Share of growth
US total	100	100	100	100
Northeast	29	11	25	10
North Central	28	18	28	22
South	24	42	31	38
West	19	29	17	30

Source: Armington *et al.* 1983.

Table 4.22. *Changes in high-technology employment for selected*
states, 1975–9 (in thousands)

	Employment 1979	Employment change 1975–9	% change 1975–9
California	574.9	154.3	36.7
Massachusetts	222.0	54.4	32.5
Texas	143.6	48.0	50.2
Florida	98.3	37.4	61.4
New York	375.0	32.7	9.6
Minnesota	104.8	29.0	38.3
North Carolina	83.7	28.7	52.2
Arizona	57.8	20.5	55.0
Colorado	53.1	20.1	61.0
Michigan	92.3	18.6	25.2
New Hampshire	36.5	16.1	79.0
New Jersey	182.2	15.2	9.1
Connecticut	94.4	14.4	18.0
Pennsylvania	209.9	13.3	6.8
Ohio	161.9	13.0	8.7
Illinois	242.5	12.6	5.5

Source: US Congress, Joint Economic Committee 1982.

however, total employment climbed by 5 percent. Much of the
employment growth has been in financial services, real estate and construction. Manufacturing employment has continued to decline
(*Business Week* 1984: July 23). The point is that New York remains
a critical 'nerve-center' within the American economy. The growing
concentration of financial services in New York is occurring despite
the 'revolution' in computerized data-processing and communi-

Table 4.23. *Regional distribution of the headquarters of
the 500 largest industrial
corporations in the United
States, 1959 and 1980*

	1959	1980
New England	19	52
Middle Atlantic	214	154
East North Central	148	129
South Atlantic	23	34
East South Central	4	16
West North Central	15	33
Mountain	6	8
Pacific	39	54
Other	32	20
Total	500	500

Source: Fortune, July 1981.

cations during the past ten years. The revolution was supposed to
decentralize business activity because headquarters' operations would
be able to locate anywhere. During the 1970s this did occur and many
corporations left the city. But for whatever reason, perhaps the
importance to financial executives of personal contact when conduct-
ing big, complicated deals, this trend appears to have reversed
(*Business Week* 1984: July 23).

The Northeast as a whole has lost its overwhelming predominance
as the center for corporate headquarters (see Table 4.23). But it still
maintains a clear advantage over other regions. The decision-making
function in American business and finance remains overwhelmingly
concentrated in the Northeast, so the relative decline in manufacturing has not meant a loss of economic and political dominance. Indeed,
one can argue that the improvement in communications and transportation since the 1960s has allowed the development of a new
round of 'internal colonialism' as manufacturing activity has
increased in the periphery relative to the core. Elazar (1968: 68–9), for
example, argues that long-distance control over branch managers and
production schedules has wiped out what little indigenous control the
periphery once exercised over its own development. Making the same
point, Cohen R. B. (1981: 211) writes:

the rise of the Sunbelt, like the development of a third world nation, can be
viewed as a new region's integration into the world as well as the national

economy. And like a dependent nation, much of the Sunbelt's industry and a significant portion of its finances remain under the control of outside economic actors.

However, it is difficult to accept that the distribution of the command centers in the American economy will not be unaffected by the regional shifts of the past twenty years. Historically, the American core was characterized by four major attributes: dominance in banking and capital markets, a leading role in international investment and export–import trade, possession of most of the headquarters of the largest corporations, and the lion's share of national manufacturing and heavy industrial capacity. A number of cities and associated hinterlands outside the present core, if not the periphery as a whole, now display features that might ultimately lead to a new 'fragmented' core for the US economy. One of these centers is San Francisco, which possesses all the major 'core' characteristics of leading northeastern centers such as New York and Boston. It has the advantage of integration into the rapidly growing Pacific Basin of the world-economy. It is also near the famous high-tech utopia 'Silicon Valley' in Santa Clara County, California, where illegal immigrants and other lower-paid workers assemble the chips and assembly boards that go into such products as computers, missile guidance systems and electronic surveillance devices (Saxenian 1984; 1985). Silicon Valley is seen by some commentators as a vision into the American future (e.g. Rogers 1985). Imitation Silicon Valleys are springing up all over the United States, often with massive public subsidies of one sort or another. But Japanese competition in this sector is intense and in 1984–5 there was a slump in demand for many of Silicon Valley's high-tech products, with many companies failing and laying off workers or moving many of their labor-intensive operations to other parts of the United States or abroad (Lindsey 1985). Other possible leading centers are Los Angeles, Houston and Miami, although the latter two lack the financial resources of Los Angeles and San Francisco. All, again, have important overseas connections: Los Angeles to Asia and Latin America, Houston to Latin America and the Middle East, and Miami to Latin America (Bensel 1984: 313).

The major long-term decline within the American economy, then, is in the heavy manufacturing industries, particularly steel and automobile production. This decline may not continue if the United States returns to a high-tariff policy that insulates the affected industries from overseas competition. Unlike the protectionism of the nineteenth and early twentieth centuries which stimulated the growth of innovative industries, a high tariff today would protect those regions most dependent on traditional manufacturing at the expense of those industries and regions moving into new 'high-tech' activities.

If present economic trends continue, a number of political consequences should therefore ensue. The most important *long-term* one from a historical perspective would be a growing divergence between the low-technology, high-wage, heavy-industrial region of the Midwest (the West of the *antebellum* period) and a resurgent high-tech, finance-centered Northeast with 'outliers' in places such as Los Angeles, San Francisco, Denver, Dallas, Houston and Miami. One major difference would be over tariff barriers and other protectionist measures. In an historical irony of some significance, the Midwest might find common cause with some parts of the South, especially the textile-producing areas, in attempts at protecting manufacturing employment. Thus would be reversed the nineteenth-century pattern of a South committed to free trade facing a Northeast committed to protectionism!

In the *short run*, however, and as a consequence of both political and economic changes, the cohesion of the political representatives from the industrial core should be maintained. This cohesion is a reflection of the present common interests of the Northeast and Midwest in redirecting federal spending and protecting heavy industry. In 1976 members of Congress from the industrial core formed an organization called the 'Northeast–Midwest Economic Advancement Coalition' to represent the interests of the industrial core in Congress. The coalition contains both Republicans and Democrats. In votes on a wide range of issues, from aid to New York City to defense expenditures, energy policy, federal aid to education, environmental policy, and distribution formulae for welfare programs, the two major parties have been badly divided between core and periphery delegations (Bensel 1984). Neither party can successfully bridge the core–periphery division of the country. Sectional conflict has re-emerged with a vengeance.

An important long-term political consequence of recent regional economic trends involves the reapportionment of seats in the US House of Representatives away from the core to the periphery. With shifts in population consequent to shifts in capital investment came shifts in political representation (see Table 4.24). A majority of members of Congress in 1960 came from the Northeast and North Central regions, roughly the industrial core. But in 1980 the South and West obtained a net advantage. Quite what they will do with this depends upon the divisions within their congressional delegations over tariff

Table 4.24. *Numbers and net changes in the number of US Representatives by region, 1910–80*

	1910	1930	1940	1950	1960	1970	1980
Northeast	123	122 (−1)	120 (−2)	115 (−5)	108 (−7)	104 (−4)	95 (−9)
North Central	143	137 (−6)	131 (−6)	129 (−2)	125 (−4)	121 (−4)	113 (−8)
South	136	133 (−3)	135 (+2)	134 (−1)	133 (−1)	134 (+1)	142 (+8)
West	33	43 (+10)	49 (+6)	57 (+8)	69 (+12)	76 (+7)	85 (+9)

Source: Calavita 1983: 75.

and other questions. Beginning with the 1984 elections, however, a political party based on a 'be kind to business' philosophy such as the Republicans, and a presidential candidate with a reputation for extreme conservatism, could have conceded the entire industrial core and still controlled the US House of Representatives, and reached the White House, solely with votes from the periphery. At least in national electoral politics the core is core no more.

Indeed, the 1984 presidential election was a competition between Democratic support in the industrial core, central cities and ethnic enclaves and Republican strength in a growing and expansive periphery, which includes the South, the West and the suburbs of major metropolitan areas (Archer *et al.* 1985). This is an exact reversal of the political pattern that prevailed in the years when the Republican party was the party of the core and the Democratic party was the party of the periphery (1896–1932). Perhaps the Republicans know something about the future of the core the rest of us do not!

A final political consequence of recent regional shifts in the United States concerns the shock administered to the independent political power of organized labor and the future of the American 'welfare state.' Organized labor has never been strong in the United States relative to most other industrial countries (see Chapter 2). The regional shifts of the past twenty years have dealt organized labor a new series of serious blows. The strength of the largest unions has been in the manufacturing industries now most in decline such as steel and automobile production. Many members of previously large and politically influential unions such as the Union of Auto Workers (UAW) have become unemployed. The new jobs are in industries and in regions where unions have had great difficulty organizing and are faced with serious legal, institutional and cultural constraints (see Figure 4.15). The future of organized labor as a force in American life depends upon its ability to turn the new circumstances to its advantage. At the moment the prospects look bleak.

The welfare state, to the extent that one exists in the United States, refers to government provision of certain basic public goods and services. As US industries have moved overseas and moved within the United States from places in the core to places in the periphery, they have been moving away from regions in which there is a political commitment to a government role in welfare services to one in which there is not. This reduces the resources available to finance such services and, ultimately, can lead to the abandonment of provision. Moreover, some regions have attracted industries by *limiting* the public services that are available. Increased industrial mobility, therefore, threatens both the fiscal base upon which welfare services can be provided and their extension to new places. Who dares to provide *public* services when the tax base may move?

A major conclusion *vis à vis* the question of the decline of the core

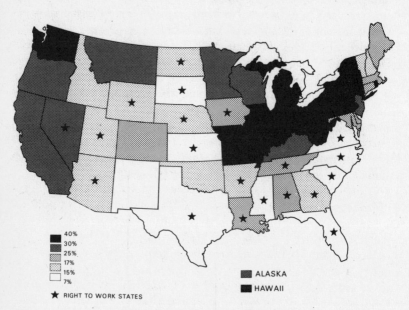

Figure 4.15: Average union membership 1970–8: the proportion of employees in non-agricultural establishments who are union members, by state (Peet 1983: 123). Reproduced by permission of *Economic Geography*.

is that the identity between the American manufacturing belt and the American political–economic core is fading. In particular, the core is increasingly *fragmented* between those regions most dependent on manufacturing industry and now exposed to the challenge of the world-economy and those which retain their hegemonic roles. But the core as a whole is also threatened by the rise of new, powerful and competing economic centers and the shifts in political power that have followed from the regional shifts in economic activity and population. The core may not yet be a new periphery but it certainly is not what it once was.

Five vignettes

The regional geography of the American impasse is incomplete without reference to impacts on specific places. The regional restructuring of the American economy has left some places with high unemployment, high rates of welfare dependency and declining incomes, while other places have become 'boom towns' with rapidly growing industries and low unemployment, if not without problems of their own in the form of high levels of pollution, traffic congestion and inadequate public services. The economic geography of the United States is increasingly local rather than sectional in structure. The grand regions of earlier periods in American history show increasing signs of disintegration. Consequently, the vignettes should not be seen simply as 'case studies' of regional experience.

But the five vignettes or 'snapshots' are *representative* of the current state of different American regions. Detroit represents in clear focus the problems and dilemmas of many industrial centers in the Northeast and Midwest. Lowell, Massachusetts, represents the possibilities for revitalization within the same regions, especially the Northeast. Audubon County, Iowa, is a farming community deep in crisis. On a scale not seen since the Great Depression, farmers in the self-styled 'breadbasket of the world' are going out of business in large numbers. Both Houston and Miami are current success stories. They are in the fast track of growth in the American economy. They are the 'new' periphery.

Detroit

Detroit is synonymous with the US automobile industry. For many journalists 'Detroit' is indeed shorthand for that industry. But in the past 25 years Detroit has lost 27 percent of its population and 70 percent of its jobs in manufacturing. By 1981 over 400,000 Detroit residents, one in every three, was receiving some form of public assistance (Luria and Russell 1984). Detroit's past growth was set by the cycle of automobile production. Today the global reorganization of the automobile industry places the 'motor city' at the heart of the crisis facing American manufacturing industry and the communities in which it is located.

In1980–1 the US automobile industry experienced its worst downturn since the 1930s. The Big Three companies (General Motors, Ford and Chrysler) lost a combined $3.5 billion, 250,000 workers were given indefinite lay-offs, and 450,000 more lost their jobs in industries supplying the major companies (Hill 1984). This crisis has major multiplier effects because the automobile looms large in the US economy. As of 1981 more than 4 million jobs directly depended on it, including 800,000 in automobile manufacturing, nearly 1½ million in supplier industries, and more than 2½ million in sales and servicing. Automobile manufacturing consumed 60 percent of the country's synthetic rubber production, 30 percent of its metal castings, 20 percent of its steel, and 11 percent of its primary aluminum (US Department of Transportation 1981).

However, the cold winds of global competition have brought a big chill to the US industry. In 1980 imports of automotive parts, engines and vehicles exceeded exports by more than $11 billion. On top of an overall decline in its world-wide competitive position the US industry recorded sharp drops in output and employment throughout the period 1977–82 (Trachte and Ross 1985).

The industry's recent contraction is by no means attributable to foreign competition alone. American companies have themselves shifted major elements of production abroad and failed to adapt to higher gasoline prices by building smaller, more fuel-efficient cars. As a result of growing international specialization, international trade in automobiles and automotive parts has grown rapidly. The growth in this trade and the surge in demand for more fuel-efficient cars have given rise to the concept of the 'world car' designed to be sold widely throughout the world. Such a car would be produced as well as consumed globally. One example of a car that is assembled from parts made in many nations is Ford's European Escort. Assembly takes place in Britain, West Germany and Portugal from components made in 17 different countries (see Table 4.25).

Detroit is a victim of the world car. Its problems, therefore, are not cyclical or temporary and not likely to be resolved through a general upturn in the US economy. The strategies that the US automobile companies have adopted in response to changes in the world market – downsizing, computer-aided manufacturing, foreign sourcing and overseas production – are exacerbating rather than relieving the economic malaise of Detroit (Trachte and Ross 1985).

The city is increasingly isolated from its surrounding area. It has suffered relatively more from the decline of the US auto industry than its suburbs or other places where employment in vehicle production was high (see Table 4.26). The quality of life has suffered consequently. For example, Bunge has dramatically illustrated the asymmetric distribution of life chances within Detroit with his map of infant mortality in the city (Bunge and Bordessa 1975). There is a four-fold range in infant death rates between the city and its suburbs. Central-city rates are comparable with those of underdeveloped countries such as Peru or Guyana. In contrast, suburban rates are comparable to the lowest in the world. The residents of the central city breathe air that is four times as dust-concentrated and contains four times the sulphur dioxide of air in the suburbs. The central city contains much less recreation space, poorer schools and much higher crime rates (Jacoby 1972; Rose and Deskins 1980).

The history of Detroit captures in one place the history of the American manufacturing belt, especially its western part. From 1870 until 1910 Detroit was one of several multipurpose cities of the industrial belt. From 1910 until the late 1960s the automobile industry and war production made Detroit a major national industrial center (Zunz 1982). But the country's sixth largest city has now become enmeshed in what Hill (1984: 315) terms 'a web of uneven development spun first by the flow of industrial and commercial capital to the suburbs and then to the Sunbelt and, more recently, by the reorganization and decentralization of the auto industry on a global scale.'

Lowell, Massachusetts

New, so-called 'high-technology' industries have been touted as a solution to most of America's economic ills. It is clear, as studies suggest, that high technology is not a panacea, particularly for places such as Detroit (Browne 1983). However, there are some places that constitute 'models of reindustrialization.' One of these is Lowell, Massachusetts, described by Flynn (1984) as a 'high technology success story.'

From the late 1820s until the late 1920s Lowell was a specialized textile-manufacturing center. By 1920 the population of the city was around 112,000 and employment in the textile industry accounted for

Table 4.25. *Sources of components for Ford's 'world car': the European Escort*

Country	Components
Austria	Radiator and heater hoses, tires
Belgium	Hood-in trim
Canada	Glass, radios
Denmark	Fan belts
France	Seat pads, sealers, tires, underbody coating, weatherstrips, seat frames, heaters, master cylinders, ventilation units, hardward, steering shaft and joints. Front seat cushions, suspension bushes, hose clamps, alternators, clutch release bearings
Italy	Defroster nozzles and grills, glass, hardware lamps
Japan	DDWS washer pumps, cone and roller bearings, alternators, starters
Netherlands	Paints, tires, hardware
Norway	Tires, muffler flanges
Spain	Radiator and heater hoses, air cleaners, wiring harness, batteries, fork clutch releases, mirrors
Sweden	Hardware, exhaust downpipes, pressings, hose clamps
Switzerland	Speedometer gears, underbody coatings
United States	Wrench wheel nuts, glass, EGR valves
England, Germany	Muffler ass'y, pipe ass'y, fuel tank filler
England	Steering wheel
England, Germany	Tube ass'y steering column, lock ass'y, steering and ignition
England, France	Heater ass'y
England, Germany	Heater blower ass'y, heater control quadrant ass'y
England, Italy	Nozzle windshield defroster
England, Germany	Cable ass'y speedometer
Germany	Cable ass'y battery to starter
England, Germany	Turn signal switch ass'y, light wiper switch ass'y, headlamp ass'y bilux, lamp ass'y front turn signal

England, Italy	Lamp ass'y turn signal side, rear lamp ass'y (inc. fog lamp), rear lamp ass'y
England, Germany	Weatherstrip door opening, main wire ass'y tires, battery, windshield glass, back window glass, door window glass, constant velocity joints
France, Germany	Transmission cases, clutch cases
England, Germany	Rear wheel spindles
Germany	Front wheel knuckly
England, Germany	Front disc
England, France, Italy	Cylinder head
England, Germany	Distributor
United States	Hydraulic tappet
England, Germany	Rocker arm
England	Oil pump
Germany	Pistons
England	Intake manifold
England, Germany	Clutch
Germany	Cylinder head gasket
England, Germany, Sweden	Cylinder bolt
N. Ireland, Italy	Carburetors
England	Flywheel ring gear

Note: Steel (body and forging barstock) from UK, Germany, Belgium, France, Italy, Austria (sheet) and Finland (bar).
Source: Fieleke 1981a: 43.

Table 4.26. *Employment in motor vehicle production: United States, Michigan and Detroit metropolitan area, 1949–82 (in thousands)*

	U.S. Total employment	Michigan Total employment	Detroit Total employment
1949	751.3	na	313
1950	816.2	na	334
1951	833.3	na	339
1952	777.5	na	315
1953	917.3	na	365
1954	765.7	na	288
1955	891.2	na	313
1956	792.5	411.7	252
1957	769.3	387.7	252
1958	606.5	288.4	183
1959	692.3	303.4	194
1960	724.1	311.2	198
1961	632.3	268.4	168
1962	691.7	299.2	177
1963	741.3	315.8	192
1964	752.7	332.3	208
1965	842.7	364.5	227
1966	861.6	381.9	236
1967	815.8	361.8	219
1968	873.7	382.8	233
1969	911.4	398.7	246
1970	799.0	333.6	210
1971	848.5	351.8	216
1972	874.8	367.7	224
1973	976.5	399.0	248
1974	907.7	364.3	232
1975	792.4	327.9	199
1976	881.0	352.8	221
1977	947.3	386.5	239
1978	1,004.9	409.6	252
1979	990.4	392.7	232
1980	788.8	333.5	192
1981	783.9	319.4	178
1982	704.8	287.0	167

Source: Trachte and Ross 1985: 210.

over 40 percent of all manufacturing employment in Lowell (Flynn 1984: 40). Between 1924 and 1932 employment was cut in half by the drop in demand brought on by the Great Depression and the movement of many firms to the South to take advantage of low wage rates and a more docile labor force. When the national economy recovered from depression Lowell did not. Economic stagnation and high unemployment characterized the area for the next thirty years (see Figure 4.16).

Beginning in the late 1960s the Lowell economy underwent a considerable revival. Growth in employment was especially rapid after the recession of 1974–5, growing by over 6 percent per year on average from 1976 until 1982. This was a rate more than double that of Massachusetts and the country as a whole (Flynn 1984: 40). Unemployment since 1979 has been around 4 to 5 percent compared to rates of 5.8 percent in Massachusetts and 8.1 percent in the United States as a whole in March 1984.

Much of the growth in employment has been in durable-goods industries such as electrical and electronic equipment, computers, instruments, and guided missiles and space vehicles. These are 'high-tech' industries. 24 percent of all employment in the Lowell area in 1982 was in so-called high-tech activities. Less than 4 percent of all employment in the United States and less than 10 percent of total employment in Massachusetts fitted the same definition (Flynn 1984: 41–2).

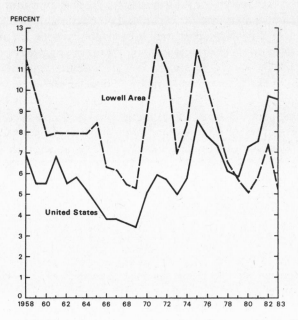

Figure 4.16: Unemployment rates in the Lowell area and in the United States, 1958–83 (Flynn 1984: 41). Reproduced from the *New England Economic Review* (Federal Reserve Bank of Boston) with permission.

Three industries – office, accounting and computing machines, guided missiles and space vehicles, and electronic components and accessories – account for 90 percent of the high-technology growth from 1976 to 1982, the first of these alone accounting for over 70 percent of the growth in high-technology manufacturing employment and 45 percent of total local job growth. There is, therefore, a vulnerability to the needs and demands of a single industry. Lowell need only look to its own past to see the dangers of this (Flynn 1984: 42).

Professional and technical occupations account for a larger share of jobs in the high-tech industries than in the 'traditional' industries of Lowell (clothing, leather, textiles and food products). But unskilled and clerical jobs account for the highest proportion of workers in the high-tech industries. Production workers are also relatively low-paid. Indeed the major expansion of jobs has been in industries in which Lowell's production wages are low relative to national averages (Flynn 1984: 43). The social structure produced by high-tech industries in Lowell is an 'hour-glass' shape, with a concentration of professional–technical workers at the top and a concentration – much larger, of course – of production workers at the bottom. There are few of the skilled craft and repair jobs that dominate in such traditional industries as automobile production or steel production.

The secret of Lowell's success has been the characteristics of its labor force. Not only did the city have access to the engineering and technical workers available in the Boston metropolitan area; it also provided a pool of relatively low-cost, low-skilled labor. Many of the places seeking to 'industrialize' their economies today do not share these characteristics. Detroit, for example, does not have access to large numbers of technical and engineering workers. Moreover, its production workers have much higher relative wages. Those still employed in automobile production are relatively highly paid. Job shifts could, therefore, mean significant wage cuts. Finally, production workers in the automobile industry are highly skilled and highly unionized. High-tech industries find neither characteristic appealing when searching for places in which to locate (Flynn 1984: 46–7).

But in Lowell a long-term tide of stagnation and decline has been reversed. High-tech industries have worked to provide the city with a new economic base. The possibilities of this working elsewhere are limited. As Flynn (1984: 47) points out: 'high-tech industries' do not have enough jobs to turn around all the depressed areas of the country.' Neither are they interested in locating in places where they must share power, and profits, with unions and workers.

Audubon County, Iowa

After California, Iowa is the premier agricultural state in the United States. With farm income of over $10 billion in 1980, only California ($13.5 billion) exceeded it and only Texas ($9 billion] came close. Iowa's major commercial crops are cattle, corn, hogs and soybeans, in that order. But agriculture in Iowa is in deep trouble.

To the naked eye much of Iowa is doing just fine. Many farms are large and in a healthy financial state. The land is so black it almost hurts your eyes to look at it. The tragedy unfolding is not visible in the landscape as yet.

Statistics cannot tell this sad story but there are a number of chilling ones (Hendrickson 1985): (1) One-third of all Iowa farmers are facing foreclosure. Their debt-to-asset ratio is of the order 70:30 or higher. (2) According to a poll in *Farm Journal* in the winter of 1984–5, 42 percent of Iowa farmers are thought to be 'sliding towards insolvency.' (3) According to a sociologist at the University of Missouri, the suicide rate among mid-western farmers is 30 to 40 percent above the national non-farm rate and rising.

In Audubon County, about 100 miles west of the state capital, Des Moines, one medium-sized family farm offers a good example of what has been happening. Six years ago Elmer and Patsy Steffes had net assets of $600,000. Their 430-acre farm had 100 milking cows, 400 sheep, pigs and, of course, corn and soy beans – a good range of products. In the 1970s they used their land as collateral to borrow and expand production. In this they were encouraged by banks and a government in Washington anxious to stimulate agricultural exports. But in 1985 the land is valued at only half of what it was. The farm is officially bankrupt. $168,081.92 at 14 percent interest is owed to the Landman's Bank of Audubon. On February 20, 1984 the bank foreclosed on the Steffeses, hoping to auction off livestock and farm machinery to get back some of its money (White 1985).

The crisis of which the Audubon County case is a manifestation has been creeping up on America's family farmers for many years. Where there were 6.4 million farms in 1940 there were only 2.7 million in 1980. Of these, under 600,000 farms, a scant 22 percent, produced four-fifths of farm output. Many of these are owned by large companies rather than family farmers. In the 1970s many family farmers overextended themselves at a time when the instability of relative profit rates, government subsidies and taxes were encouraging large companies and banks to diversify into agriculture. About one-third of American agricultural acreage is now owned by absentee landlords. The real crisis, however, came between 1979 and 1981, when as a result of the rising dollar and the stagnation of export markets farm income dropped by nearly one-third in real terms. Good harvests reinforced the problem, increasing the cost of government price supports yet driving down prices to below average costs of production (Harris 1983: 211; Robbins 1985a; Benjamin 1985).

If present trends continue, the family farms and small towns of Audubon County are finished (Robbins 1985b). The large agribusiness and food production corporations may well take over, as they have already in California. Audubon County will then be neither socially nor environmentally viable by 2020. Michael White (1985: 9) puts it another way:

If one bank takes the Steffeses cattle and tractors, another bank will have no choice but to take their land. Mr Steffes assumes it would pull down his farmhouse and outbuildings and manage the farm as part of a large unit. The Steffeses, like many others here, will end up on welfare – food producers getting food stamps.

Houston

Houston has been described as the 'preeminent Sunbelt City' (Feagin 1984: 99). It was relatively small until the 1920s. The population was only 78,000 in 1910. But it grew as the Texas oil industry grew. By 1940 its population was 385,000. Since World War II the growth has been even more spectacular. In 1980 the population was 1.7 million for the city proper with an additional 1 million in the greater Houston area. From 1970 to 1980 Houston was the fastest growing of the 15 largest metropolitan areas in the United States (Feagin 1984: 103).

Houston today is a decentralized city sprawling over approximately 1,000 square miles of the Texas Gulf Coast. There are seven major business activity centers: the downtown area, the airport area (north), three centers in the southwest corridor out from the center, and two centers in the northwest. Scattered between these commercial–industrial nuclei are residential areas, including condominium apartment buildings and sprawling suburban subdivisions.

The city is headquarters to the American petrochemicals industry. Most of the country's oil and gas companies have offices and facilities in the greater Houston area. About one-quarter of US oil-refining capacity and one-quarter of the oil–gas transmission companies are located in the Houston and Gulf Coast region. The flow of profits from the oil–petrochemical industrial sector has provided much of the capital that has expanded Houston industrially and spatially. In turn this solid base of local investment capital has stimulated investment from elsewhere in the United States and overseas (Feagin 1984: 101).

Since 1970 more than 200 corporations, many of them not in the oil business, have opened major facilities, including headquarters, in Houston. Houston is not only the world's major oil and gas center but also an important national center for such manufacturing industries as forest products, for agribusiness and for banking.

Behind the growth lies an image and a reality cultivated by the city's established business leaders: that of Houston's 'good business climate,' a euphemism for an area with lower wages, weak unions, lower taxes and a conservative political culture. Undoubtedly, Houston's lower production costs (cheaper energy, weak unions, lower wages) and limited barriers to new development (no old industries or aging infrastructure) have, along with helpful federal expenditures on highways and defense industries, stimulated much of its recent growth.

But local government has also been important. Feagin (1984: 115) again has a good phrase for it: 'the business of Houston's government has from the very beginning been business.' The city government has limited its regulation of businesses to a minimum, provided services as requested by the business community, and poured tax dollars into public infrastructure – sewers, streets, bridges, etc. Regulation has been particularly weak. For example, Houston has no zoning regulations. As a result there is little government planning of land-use change. Houston is the ultimate 'free enterprise' city. If you have the money you can do what you like.

However, all that glitters is not gold. There is another side to the 'Houston miracle.' In the first place, Houston, and Texas as a whole, has benefited from high energy costs. In 1985, as world oil prices declined dramatically, Houston stopped growing. Houston is as vulnerable to changes in the world-economy as anywhere else (Reinhold 1985; Weinstein and Gross 1985). Second, Houston has a huge poor- and moderate-income population. Outside of California it may have the largest concentration of illegal immigrant workers, mainly from Mexico, in the United States. There are also massive environmental problems. For example, the city is gradually sinking. At only 50 feet above sea level the city is especially vulnerable to depletion of its underlying water table, a process which weakens the supporting soil structure. When added to by oil and gas extraction and the weight of all the new development construction, the soil drainage of the past 70 years has led to a drop in land elevation of several feet. Over the next several decades subsidence will create major flooding, construction and demolition problems (Feagin 1984: 119–20). More conventional environmental problems include heavy traffic congestion, air pollution and sewage discharge problems.

Houston's low taxes, not surprisingly, have meant inferior government services. The public education system is abysmal (Weinstein and Gross 1985). Water pressure is low in many areas. Provision of subsidized housing is woefully inadequate (Feagin 1985). However, the most surprising features of Houston given its boom town image are the high levels of unemployment and poverty. At around 20 percent, unemployment is as high as in many declining cities of the American Midwest and Northeast. And underemployment – part-time and low-wage full-time employment – is even more of a problem in Houston, especially in black and Chicano neighborhoods. A fifth of Houstonians live at or below the official federal 'poverty line' (Feagin 1984: 122–3). But as one commentator sees it: 'These costs don't show up on any firm's ledger; no accountant writes them down. They're not charged against the income the firm makes from selling its products and services' (Smith, D. 1979: 3). Therefore, they don't count, they don't matter.

Miami

It's one of the hottest shows on American television. *Miami Vice* is a weekly series of adventures of two cops in the murky and violent world of Miami's criminal underworld. According to some accounts it is also non-fiction. When the narcotics boom took off in the mid-1970s Miami became the drug capital of the world. More than 70 percent of the US supply of heroin, cocaine and other illegal substances flows through it. This traffic has brought drug-related crime – Miami's murder rate is the highest in the country – and wealth. The influx of 'hot dollars' has made Miami a rival to New York City as the nation's financial capital (Lernoux 1984).

Miami boomed throughout the late 1970s. Many banks and multinationals were lured to Miami by its huge cash-flow. Not all of this came directly from drug-smuggling and money-laundering. In particular, exporters, retailers and real-estate agents catered to the more than 2 million Latin American refugees and tourists who flocked to the city each year from 1975 to 1983. With its 45 percent Hispanic, predominantly Cuban, population, Miami became Latin America's playground and political refuge. Argentinians and Venezuelans bought many luxury condominiums, gambling that the city's real-estate market was a better investment than their own inflating currencies (*Business Week* 1984: April 30). For those on extended shopping trips there were cameras, watches, televisions, and designer blue jeans. If these could not be carried home they could be shipped through the convenient Port of Miami container terminal. Exports through Miami reached almost $8 billion in 1981, about 80 percent going to Latin America (*Business Week* 1984: April 30). For those interested in 'dollar salting' Miami became a substitute for Zurich. During the boom in lending by American banks to 'Latin America' in the late 1970s many American banks set up branches in Miami to capture their share of the $3 billion to $4 billion fleeing Latin America each year (*Business Week* 1984: April 30).

These infusions, legal and illegal, transformed Miami. From a city historically dependent on the pension incomes of its large population of retirees and the spending of tourists from the North flocking to sun and beaches, Miami became in ten years a Latin American Hong Kong. The prime customers became Latin tourists and Latin business.

The boom changed the orientation and the skyline of Miami as well. The center of activity moved away from the hotels on the beach to a downtown exploding with bank and other high-rise buildings. The most prestigious banking address is Brickell Avenue, with its gold-and-black skyscrapers. Coral Gables, where a hundred multinationals have their Latin American headquarters, has also become a chic location. When added to the more than 100 banks representing 25 countries, the scale of Miami's financial activities comes into focus (Lernoux 1984).

In 1982 the Latin tourist boom collapsed. The world recession of the early 1980s, the ending of big US bank loans, and the austerity packages imposed by the International Monetary Fund, have led to a devastation of the Latin economies. Exports through Miami plummeted by 27 percent ($2 billion) from 1982 to 1983. The great dream of Miami as a gateway to Latin America faded. But the banks have not moved. Only a few multinationals have left town. There has been no mass exodus. The companies that have left do not blame their departure on the Latin American collapse. General Electric, for example, decentralized its Miami operation to separate country headquarters in Mexico, Venezuela and Brazil (*Business Week* 1984: April 30). Banks and business in general are awaiting a revival of the Caribbean and South American economies.

In the meantime there are still the transfer payments – pension checks, social security, retiree saving accounts and, of course, the drug money. It will not be easy for the city to go back to dependence on retirees and American tourists. In particular, Miami's vacation and retirement image has been repeatedly tarnished by crime and inter-ethnic conflict. Despite massive federal enforcement programs and frequent 'declarations of war' on the drug traffic, Miami remains the drug-smuggling capital of the United States. Too many people, includ-ing those in respectable organizations like those banks waiting for Latin American recovery, have too much at stake in the drug business (Lernoux 1984). *Miami Vice* is good television, but it does not make for such good non-fiction.

Conclusion

In the period since the late 1960s the United States has reached an impasse in its previously hegemonic position within the world-economy. Evidence of relative decline abounds. This chapter has surveyed this evidence and related it to the changing regional geography of the United States. For the first time since the Civil War the dominance of the northern industrial core and its associated manufacturing belt is seriously in question. The macroeconomic con-text of this shift is important. Now that the overall growth rate in the American economy is lower, the regional allocation of that growth becomes even more important. Quite what will happen now remains to be seen. One consequence is clear for both Sunbelt and Snowbelt: the years of certain boom, of growth as a taken-for-granted, are past. Even for the United States, so much its spiritual home in the twentieth century, 'capitalism does not simply shed its old skin, staying in one place. Like the locust it moves on' (Harris 1983: 51).

5

From challenge to responses: the
United States versus the world-economy

On the evening of November 9, 1965, just after nightfall, the electrical supply for most of the northeastern United States failed. America's largest city, New York, was blacked out, along with its entire metropolitan area, as were the populous states of Connecticut, Massachusetts and Rhode Island. The unquestioned belief that unlimited use of electrical power was the birthright of all Americans and that electricity made individual Americans 'free' and 'independent' should have ended that night. Power did not, it would seem, make Americans free. Rather, it made them prisoners of machines, of power-grids, of corporations, of experts. Power made Americans dependent.

The following twenty years have more than reinforced this impression. A variety of 'power failures' come to mind among all the images the word 'power' elicits for Americans. As one writer puts it:

'Power failure' is a blackout or a brownout, it is a failure in expertise, it is Three Mile Island; it is also Watergate. On a much larger scale, it is what Americans now call simply 'Vietnam.' It is the inability of a large and complex corporate–military–government power elite, backed by all the wealth and might of a modern superpower, the United States of America, to free fifty Americans from a building in Teheran. (Robertson, J. O. 1980: 279)

In 1986 it was the explosion of the space shuttle 'Challenger' (Brummer 1986).

There have always been misgivings about American power, although it is only over the past twenty years that they have become more widely discussed. Ever since the early American corporations began to grow with the extension of railroads and the building of large factories there have been Americans who have had serious objections to modern industrial power and the American exercise of it. But they have been a small minority, and relatively powerless too. Much more fundamental and persistent has been a *way of life*, not mere sentiments, based upon a faith in what Williams (1981: 113) calls 'infinite progress fueled by endless growth.' Growth was the founding concept of American culture. As Tocqueville noted in 1835, 'restless-ness in the midst of abundance' was nothing new for individuals, 'the novelty is to see a whole people furnish an exemplification of it.' Americans are, he added, 'forever brooding over advantages they do not possess,' and are obsessed 'with the bootless chase of that complete felicity which is for ever on the wing' (de Tocqueville 1966: 535–8).

It is in this context that America has become a victim of its own success. Rapid growth in the past, especially in the 1950s and 1960s, has been elevated in memory to a sacred pedestal without much thought to changed circumstances. In the 1950s and, especially, from 1961 until 1966, the American economy expanded spectacularly. Incomes rose dramatically. Production soared. Unemployment dropped below 4 percent. Those were the halcyon days. Prosperity increased so fast that people could see and feel it. To many Americans this quickly came to be *normal*. When you currently hear talk about getting the economy 'back on track,' the track in mind is the fast one of those golden years of growth. Nostalgia for the 1950s is a major feature of America in the 1980s.

The great challenge to Americans is to recognize that the growth in the past represents not a sustainable trend but an extraordinary and temporary swing far above it. Wallerstein (1984: 68) puts it rather bluntly: 'Americans have spent the past thirty years getting used to the benefits of a hegemonic position, and they will have to spend the next thirty years getting used to life without them.' The critical questions then become whether they will and, if so, how?

To many commentators and, one suspects, many Americans, given the rootedness of the culture of growth, the present American impasse is a temporary and transitional phenomenon as the country shifts from an 'industrial' to a 'post-industrial' society. Under a new guise, happy growth days will soon be here again. The first section of this chapter is devoted to this conception of the present impasse. A second section takes a different point of departure. It argues that America needs not an expansion of a flawed and declining economic system but a recognition of the economic and political costs this system now entails. Post-war economic and imperial expansion obviated the need for political choice. Their collapse offers the opportunity to build a different society at home and abroad. In Wolfe's (1981a: 247) words: 'Less satiated consumers [Americans] may find themselves more satisfied citizens.'

Into post-industrial society?

To many of those who focus on the United States in isolation from the world-economy or who see the world-economy as a series of intermittent, exogenous shocks rather than its being part and parcel of American development, the United States is currently in transition from an industrial to a post-industrial society (e.g. Clark 1985). The impasse of the past twenty years is a crisis caused by this *natural* transition rather than by the changed political–economic relationship of the United States to the world-economy. To Daniel Bell (1973) there are five components of the term 'post-industrial society.' In the economic sector there is a shift from a goods-producing to a service economy; occupationally, there is a 'pre-eminence' of the professional and technical class; the axial principle of the new society is 'the centrality of theoretical knowledge as the source of innovation and of policy formulation for the society' (1973: 344); there is technological forecasting; and decision-making takes place by means of a new 'intellectual technology' (1973: 377).

Bell, one of the most widely quoted and influential proponents of the post-industrial society scenario, relates the growth of this society to the rise of the welfare state, particularly in its role as manager of the national economy. He writes (1973: 398):

A post industrial society . . . is increasingly a communal society wherein public mechanisms rather than the market become the allocators of goods, and public choice, rather than individual demand, becomes the arbiter of services . . . if the major historic turn in the last quarter of a century has been the subordination of economic function to societal goals, the political order necessarily becomes the control system of the society.

Although Bell nowhere makes the point explicitly, his assertions about the movement towards post-industrial society suggest strongly that this society is not only post-industrial but also post-capitalist. If the 'economic function' has been subordinated to 'societal' goals by means of the 'political order' representing 'a communal society' one might infer that the post-industrial society is indeed Socialist. But if the factual assertions that underpin Bell's post-industrial society are false, then the concept itself and the implication that capitalism is disappearing are too.

First of all, when the government intervenes in an economy dominated by private corporations to promote the 'public good,' the corporations will be the primary beneficiaries of that intervention. The macroeconomic planning of the welfare state since its modern incarnation in the New Deal and during World War II has followed capitalist priorities. One need look no further than housing and tax policies. The government and the 'societal' goals it can serve have been subordinated to private goals, *not the other way around*. That has been the burden of the argument and the evidence presented in earlier chapters.

Second, the so-called 'service sector' of most industrial countries *has* expanded at relatively higher rates than in the past and service employment is more important than manufacturing employment in all industrial economies (Magaziner and Reich 1982: 47). However, the growth of the service sector *never* compensates for the decline of manufacturing (*Business Week* 1986: March 3). In particular, many service jobs are immediately dependent on manufacturing rather than independent of it. They *follow* manufacturing rather than *replace* it (Pollack 1986). Much employment in service industries such as transportation, banking and insurance is of this nature (Walker 1985; *New York Times* 1985: October 7).

Many, if not most, of the new service jobs are also much lower paid than the manufacturing jobs they are replacing. Women, who now constitute 43.5 percent of the American labor force, fill many of these jobs. There is thus no straightforward substitution of service for manufacturing jobs. The new service employees are largely different people from those once employed in manufacturing jobs (Noble 1986). In addition, as Leontief (1985) points out, 'the shift of demand from commodities into services is bound to slow down, while displacement of labor by increasingly efficient machines seems to have no limits. The ability of the service sector to absorb displaced workers will diminish.' For example, word processors and personal computers have already cut into secretarial positions as managers and executives use letter-quality printers rather than secretaries for final drafts.

Moreover, manufacturing continues to be a much more dynamic multiplier of productivity and income growth than private- and public-sector service employment (Brown and Sheriff 1979). Finally, in terms of the export market, trade in services is dependent, in the long run, on a successful manufacturing base. The greatest portion of US trade in services derives from the investment income of American manufacturing firms and of wealthy individuals, and of service exports such as banking, insurance and consulting services. All of these are either directly or indirectly dependent on merchandise trade. For example, German and Japanese banks have supplanted British banks in many parts of the world as their manufacturing sectors have become relatively stronger (Magaziner and Reich 1982: 85–6). A simple switch from manufacturing to services entails significant costs that cannot be glossed over by assertions such as:

material needs largely have been satisfied thanks to increased work-productivity, as a result of the Industrial Revolution (in which steam power was applied to industrial settings). Humankind can now turn to the satisfaction of other less material or basic needs, especially those in which information is a main component. (Rogers 1985: 3).

This is poppycock; there is no evidence for it whatsoever.

Third, proponents of the post-industrial society scenario assert that a 'new class' of scientists, economists, engineers and inventors of the 'new intellectual technology' are becoming both a major presence and a *controlling* element in modern American society. The first of these assertions is palpably false. The most recent projections of the American Bureau of Labor Statistics show that demand for such occupations as computer system analysts, engineers and computer programmers is relatively minor compared to that for such jobs as janitors, nurses' aides, salesclerks and secretaries (Riche *et al.* 1983; *Business Week* 1986: March 3; Thurow, 1986). These are the service jobs of the future (see Table 5.1). The term 'services' is on this evidence a 'chaotic' conception (Sayer 1984: 127). Lumping together hairdressers, computer consultants and plumbers is not very enlightening, yet it is central to the conception of the 'service economy' in post-industrial society.

Furthermore, to what extent is there a new technocratic–managerial class in control? Bell (1973: 31) tells us that 'the university, which once merely reflected the status system of the society, has now become the arbiter of class position. As the gatekeeper, it has gained a quasi-monopoly [*sic*] in determining the future stratification of society.' This is quite wrong. Numerous studies have found that academic achievement is *not* a good predictor of adult social achievement (e.g. Moynihan and Mosteller 1972: and Cohen, D. 1972). In the 1970s there was an 'oversupply' of educated labour which led to the employment of many college graduates in jobs for which only high-school training had previously been required (e.g. as taxi drivers or as secretaries). As Cohen (1972) put it: 'nowhere can we find any empirical support for the idea that brains are becoming increasingly more important to status in America.' Rather, the reverse is more likely. Moynihan and Mosteller (1972: 22) summarize as follows: 'family background, measured in social class terms – primarily education of parents, but including many other considerations, such as the presence of an encyclopaedia in the house – is apparently a major determinant of educational achievement.'

What happens to the 'brainy' ones when the leave university? Do they obtain jobs for which they are specially prepared? Some do become experts and decision-makers. But if they are replacing the capitalists – the entrepreneurs, the businessmen and the industrial executives – are they operating a new calculus, observably different from the old one? Although there are new recruits to business with biology degrees and computer science diplomas their business is business as usual. The bottom line on the balance sheet has not yet been replaced by an alternative metric of corporate performance. So much for the 'centrality of theoretical knowledge!'

Fourth, and finally, a major claim made for post-industrial society is that it is 'increasingly a communal society.' This claim rests on a systematic exaggeration of the amount and permanence of the social innovation that went on in America in the 1960s. To be fair, this is more obvious in 1986 than it was in the early 1970s when most arguments for the post-industrial society were first proposed. In the early 1970s issues such as school desegregation and busing to maintain racial balance in schools were controversial and symbolic of a new judicial and executive activism that seemed to portend a new 'communitarianism.' It is quite clear now that this era has passed. The mobilization of the American right in the late 1970s represents an important backlash against the limited movement towards a widely shared notion of a collective social conscience and a caring, compassionate government.

Much of this new mobilization is concerned with cultural and familial issues but has spread from these to encompass economic and foreign-policy positions as well. As Petchesky emphasizes: 'The politics of the family, sexuality, and reproduction – and most directly, of abortion – became a primary vehicle through which right-wing politicians achieved their ascent to state power in the late 1970s and the 1980 elections' (quoted in Wolfe 1981b: 14). Right-wing attacks on abortion, and the US Supreme Court for legalizing it in 1972, involve above all a reconceptualization of the private and the public, but in terms of the personhood of the foetus rather than the interests of the community.

In fact, however, the forces against which the American right are reacting are the products of capitalist development rather than a

Table 5.1. *The myth of high-tech jobs*

Occupation	New jobs by 1990
Secretaries	700,000
Nurses' aides and orderlies	508,000
Janitors	501,000
Salesclerks	479,000
Cashiers	452,000
Nurses (professional)	437,000
Truck drivers	415,000
Food service workers	400,000
General office clerks	277,000
Waiters and waitresses	360,000
Stock clerks	262,000
Elementary school teachers	251,000
Kitchen helpers	231,000
Accounts and auditors	221,000
Helpers (trades)	212,000
Automotive mechanics	206,000
Blue-collar worker supervisors	206,000
Typists	187,000
Licensed practical nurses	185,000
Carpenters	173,000
Bookkeepers	167,000
Guards and doorkeepers	153,000
Computer systems analysts	139,000
Store managers	139,000
Physicians	135,000
Maintenance repairers	134,000
Computer operators	132,000
Child care workers	125,000
Welders and flamecutters	123,000
Electrical engineers	115,000
Computer programmers	112,000

Note: The table is based on Bureau of Labor Statistics projections and shows those occupations that will provide the greaters number of new jobs by 1990.
Source: Harper's August 1984: 24. Taken from *The New Manufacturing: America's Race to Automate.*

'communal society' produced by 1960s permissiveness. That is something of a red herring in this context! Inflation, working spouses, 'latch-key' children, pornography, crime, drug abuse, environmental degradation all have some relationship to the perennial search for new sources of capital accumulation. Patriarchy, fundamentalist Christianity, creationism and political authoritarianism are the responses when a more imaginative and less punitive repertoire is not culturally validated. Indeed, much of contemporary American culture seems pre- rather than post-industrial in nature!

The claim, therefore, that America is moving into a post-industrial society and that this move is apparently costless, indeed beneficial, especially in the long run, is false. The evidence, rather, is that the United States is a declining industrial society exhibiting cultural and political, as well as economic, symptoms of its decline. This could be called 'post-industrialism' but it would mean something quite different from what Bell and his followers have in mind.

From impasse to political choice?

From this alternative or antithetical point of view the prospects for the resurgence of the American economy and the hegemony it has enjoyed within the world-economy are dim. Growth, therefore, can no longer substitute for political choice. It is in this context that alternative prescriptions for coping with the impasse must be considered (Balbus 1982).

Of course, those dominant politically share neither this prognosis nor prescriptions based upon it (Connolly 1983). They continue to believe that growth and empire are still possible. Thus the Republican Party platform in 1980 said the following about economic growth in 1980:

The Republican Party believes nothing is more important to our nation's defense and social well-being than economic growth . . . With this kind of economic growth, incomes would be substantially higher and jobs would be plentiful. Federal revenues would be high enough to provide for a balanced budget; adequate funding of health, education, and social spending; and unquestioned military preeminence, with enough left over to reduce payroll and income taxes of American workers and retirees. Economic growth would generate price stability as the expanding economy eliminated budget deficits and avoided pressure on the Federal Reserve to create more money. And the social gains from economic growth would be enormous. (Republican Party 1980: 26)

Without growth, objectives such as balancing the budget and increased military spending are mutually incompatible. Rather than admit that the prospects for such growth are much diminished compared with the 1950s and early 1960s and so pose the hard alternatives needed to release America from its impasse, the major political parties make competing claims about how to achieve objectives that can no longer be achieved. Unless the economy is dramatically restructured and military spending reduced, the impasse will deepen rather than lessen.

One major lacuna in the growth-at-any-price approach to national politics concerns consideration of the question: growth for what? During the growth era after World War II public goals were uncontroversial and removed from public discussion – to win the arms race with the Soviet Union and beat 'them' to the moon; to maintain domestic economic growth by 'fine-tuning' the federal budget and the money supply; to allow private business free rein to accumulate capital in the 'national interest.' Political debate was replaced by competing claims concerning managerial efficiency. 'Correct' decisions could be reached technocratically – by hitting the right velocity for the money supply or creating tax incentives. As long as growth continued, and most people got something, political choices could be avoided (Reich 1983).

But there is no 'best' solution to how the costs and benefits of declining growth can be allocated socially and geographically (Harrington 1986). There is a wide range of solutions. They are only discoverable and can only be legitimized through the process of political debate and choice. Civic virtue takes on a premium in a no- or limited-growth economy. When people cooperate, the collective power of their talents and resources is greater than the mere sum of their individual contributions (Wilber and Jameson 1983: 238–41). But civic virtue is currently undervalued in American society. The ideology of a market economy in which people are motivated solely by greed and fear is dominant. Reich (1983: 270) argues that this outlook is also crippling the US economy:

A society that simultaneously offers its members both the prospect of substantial wealth and the threat of severe poverty will no doubt inspire occasional feats of dazzling entrepreneurialism. But just as surely, it will reduce the capacity of its members to work together to a common end and to adapt themselves collectively to new conditions. The ideology of wealth and poverty, to which some Americans still cling, is suited to a simple frontier economy in which social progress depends on personal daring. It may well have been a fitting ideology for America's early era of mobilization. But a social ethos shaped in a virgin continent is the wrong vision to guide and motivate the members of an increasingly complex industrial economy.

Economists and most politicians still refuse to treat the economy as a political institution in which relative power determines relative prosperity. However, a number of proposals have been put forward that, while accepting the reality of America's present impasse, see it as an opportunity rather than a tragedy, an opportunity to question the very features of the US economy that brought growth in the 1950s and 1960s – economic concentration, growing government intervention, and expansion overseas – but became liabilities in the late 1960s. They build on a serious discussion of the relative strengths of markets and states that has emerged since the late 1960s (Rawls 1971; Nozick 1974; Lindblom 1978). They also share a common democratic outlook. This is captured by Kolakowski (1978: 529) as follows: 'To create institutions which can gradually reduce the subordination of production to profit, do away with poverty, diminish inequality, remove social barriers to educational opportunities, and minimize the threat to democratic liberties from state bureaucracies and the seductions of totalitarianism.'

Some of the most concrete proposals have come from the International Association of Machinists and Aerospace Workers (1984) and the Black Coalition for 1984 (1984). Both demand comprehensive national economic and social planning, limitations on military spending, and democratic participation in economic decision-making. They both also contain legislative programs and deal with the impoverishment of political life in the face of the refusal to confront the political choices implicit in economic policy. Each work echoes the thinking of such intellectual pioneers as Wassily Leontief, Barry Bluestone, David Gordon and Martin Carnoy on the economy; Richard Barnet, Walter Lafeber and William Appleman Williams on American foreign policy; Frances Fox Piven on poverty; and C. Wright Mills on American society.

The most comprehensive blueprint for 'rebuilding America' is *Beyond the Wasteland* by Bowles, Gordon and Weisskopf (1984). This attributes recent American economic decline to many of the institutional developments described in detail in earlier chapters. The authors propose an economic program for redirecting economic energy away from such 'wasteful' practices as advertising, military spending and excessive energy consumption. Though this clearly represents a political judgment, so would a 'neutral' economic model that treats every unit of account as equivalent to all others. A connection is made with the themes of numerous other recent studies (e.g. Carnoy and Shearer 1980; Alperovitz and Faux 1981; Cohen, J. and Rogers 1983; and Howe 1984). In particular, the American economy as currently constituted is viewed as grossly inefficient. It produces vast quantities of goods for which demand must be created and visits destruction on communities in the name of progress.

The theme of 'economic democracy' is especially strong in recent writing on bringing politics into the American economy. Economic democracy is an egalitarian form of political–economic structure in which a serious attempt is made to democratize the economic sphere in general and workplaces in particular. In this way the gains and losses of economic change are more equally distributed across social strata and regions than under alternative arrangements. Economic democracy differs from democratic capitalism (in both *laissez-faire* and welfare-state manifestations), in which democracy is limited to periodic involvement in electoral politics and the means of production are largely privately owned. It also differs from conventional socialism, especially of the Soviet variety, in which markets are prohibited (in public), there is little meaningful electoral politics, and the means of production are owned by the state.

Calls for the economic democratization of American society focus on three 'spheres,' to use Walzer's (1983) term: the workplace, control over macroeconomic policy and investment decisions, and politics. Noting the centrality of work in most people's lives, proponents of economic democracy insist that this aspect of life must be democratized. For many writers the major consideration, in addition to the relative allocation of hardship, is the fulfillment of the democratic promise at the core of the Jeffersonian strand in American political thought: people have a right to an involvement in decisions that affect them. Albert Gallatin, Treasury Secretary under Jefferson and Madison, expressed this basic idea as follows: 'The democratic principle on which this nation was founded should not be restricted to the political process but should be applied to the industrial operation as well' (quoted in Derber 1970: 6).

The revival of interest in economic democracy comes at an opportune time. This reflects consideration of practices in other countries as much as abstract theorizing. For example, a number of European countries that have in recent years equalled or surpassed American living standards require that employees be allowed extensive participation in management (Magaziner and Reich 1982). In West Germany, 'co-determination' allocates positions on corporate boards to employees. In Sweden, after much controversy, a tax on corporate profits will be democratically managed and used to purchase control of corporate assets (Lappé 1983). In addition, the success of several large-scale prototypes, notably in Yugoslavia and in the Mondragon region of Spain, suggest the possibility of efficient and democratic self-managed enterprises as an alternative to present corporatism (Tyson 1980; Thomas and Logan 1982).

The growing interest in economic democracy also reflects recent practical experience in the United States. At the heart of recent American economic trends has been the displacement of people from jobs and the places in which they live. The real issue for economic democracy concerns the fact that humans are relatively immobile compared to financial capital. By 1988, if present trends towards economic concentration continue, three hundred or so giant firms will control half of the world's goods and services. Most of these are, and will be, devoted to one objective: maximizing their net earnings. They will do this by negotiating the best deals for themselves and moving on if a deal turns sour (Goodman 1984).

Some workers faced with the imminent departure of their employers have attempted to save their jobs by buying their factories. High unemployment, coupled with the realization that technology is restructuring the American economy, has pushed many workers to consider expanding the sphere of democracy. These workers reject the calculus of business and its apologists concerning the dichotomy between 'place prosperity' and 'people prosperity.' In this dichotomy a place can be abandoned and if everyone moves elsewhere everyone is better off. Mobility is the solution to disinvestment (Agnew 1984). To many people, however, their sense of identity is bound up with their place of residence and livelihood. Numerous studies suggest the importance of local ties and social vitality for life in American working-class communities. Economic democracy, therefore, means defending place against capital as much as creating a more democratic way of life. This is its great contemporary appeal in places afflicted by disinvestment (Woodworth *et al.* 1985).

There are significant barriers to economic democracy in the United States and these should be understood. One barrier of some importance is that people may identify far too strongly with 'meritocracy' ever to accept such a system. The two dominant mentalities among Americans are the success-oriented, where success is seen in improving one's lot according to existing rules, and the leisure- or family-oriented, where a job is seen as a matter of 'putting in time' in exchange for a paycheck. Neither mentality is congenial to economic democracy.

This objection is misleading in presupposing that an individual is faced with an either/or situation. In fact, participation at work is compatible with a wide range of other interests. It need not be an all-consuming passion. Indeed, surveys suggest that many people would be more interested in their work if they had more control and influence in the workplace (Carnoy and Shearer 1980).

A more serious barrier is the resistance of those in power. Since economic democracy entails changing the existing distribution of power, those who stand to lose will fight vehemently. The forces who stand to lose are very significant. Managers, capitalists and union leaders would all have to give up power. But there are encouraging signs. For example, new technologies and employee resistance have led some managers to try a host of participatory experiments and 'quality control circles.' Many influential economists and business leaders are pushing for greater worker involvement as a prerequisite for better business performance (Magaziner and Reich 1982; Rohatyn 1984; Thurow 1980). None of these proposals is exactly what economic democrats have in mind, but they do open up the possibility of debate about key aspects of economic democracy. More economic decline may induce more debate, and action.

A final problem concerns the tension between market and plan in proposals for economic democracy. Comisso (1979) compares this tension to that between Jeffersonians and federalists. She distinguishes those who, in the tradition of Proudhon, favor a loose federation of relatively powerful and autonomous democratic firms from those who, in the tradition of the Italian Marxist Gramsci, favor a relatively powerful government that plans investment and plays a major role in regional and sectoral equalization. Comisso sees this tension as a weakness. But surely this is no more a weakness in economic democracy than centralist/decentralist emphases are a weakness in political democracy. The 'tension' is an important factor in setting the agenda for political debate over the *form* economic democracy should take, not a flaw in its conception.

Social change in the United States is notoriously slow. It took many years for slavery to end even when a large part of American society rejected it. It took another hundred years for black Americans to establish their constitutional rights in everyday practice. The Equal Rights Amendment has massive public support but cannot pass a majority of the legislatures in the American states. If the American economy is to democratize, the economic crisis will have to deepen. Each stage of democratization – advisory councils, representation on boards, co-determination, full democratization – will meet with massive and well-funded resistance. But the possibility is now there. Choices can be made.

What might happen otherwise?

But two related issues still dominate American politics: economic growth and military 'strength.' In the absence of the *structural* changes in American society discussed in the previous section, they are likely to remain at the top of the national political agenda. In particular, militarism is a central aspect of contemporary American government policy. It provides profits and some jobs, as well as channeling frustrations and insecurities against outside enemies (Nincic 1985).

The *Rambo* phenomenon symptomizes the high level of xenophobia and the continuing love affair with empire that characterize contemporary America. The movie *Rambo* starring Sylvester 'Rocky' Stallone in the title role is the most popular adults-only film ever screened (Reed 1985). As a former Green Beret, Rambo's task is to find a jungle camp for Americans missing in action from the Vietnam War. In the process he kills several hundred people using a variety of techniques. He is a 'human war machine' who blows up numerous bamboo huts, an entire village, a huge Russian helicopter, two boats, a rice paddy, and about half a battalion. He is Revenge. As one reviewer aptly puts it, he wins 'in the cinema the war the United States had lost on the ground' (Reed 1985).

The justifications for increased American military spending are unclear or fictive except in the contexts of 'power projection' and economic stimulation. The sorry saga of Soviet defeat and humiliation in every aspect of its foreign policy since 1948 – the loss of China as an ally, the open resistance to Soviet domination in Poland, the defection of Egypt and much of the Arab world, the economic bankruptcy of most of its client states such as Vietnam and Cuba – let alone its own internal woes, offers little justification for the portrayal of the Soviet Union as a successful superpower. Indeed, one can make little sense of the arguments of either Soviet or American leaders except as cynical attempts to hold or recruit allies and provide a defensive screen behind which other imperatives – economic and psychological – are operative. Harris (1983: 217) sees it this way:

While the arguments clatter on, the military machines grow. By the late seventies, the nuclear stockpile of the two largest states was sufficient to destroy every city in the world seven times over, and equal to sixteen tonnes of high explosive for every inhabitant of the planet. World military spending has increased four times over in real terms since 1945; it reached 500 billion dollars in 1980 (70 percent of this spending was by the powers of NATO and the Warsaw Pact).

There is evidence that the United States is developing new alliances. Soviet–US detente has collapsed, but so largely has NATO. Fundamental differences in political and economic interests have emerged between the United States and Western Europe. In particular, American economic policy since the 1970s has largely been at European expense (see Chapter 1). There are also major disputes over US–European trade, European–Soviet trade, and recent changes in American nuclear and other military strategies, especially the so-called Strategic Defense Initiative or 'Star Wars' (Frank 1983). Beginning with Nixon, the United States set about developing a 'Pacific Rim' strategy, based on China and Japan. But Chinese suspicions of American intentions and Japanese economic interests limit American options in this direction too. Perhaps, as Vidal (1986: 19) argues, the United States may choose to ally itself with the Soviet Union if it is to survive 'in a highly centralized Asiatic world!'

In the United States the welfare state is beginning to crumble. In the context of increased military spending and economic stagnation this is not surprising. Given the imperatives of economic growth and military strength, something must be sacrificed. The 'fiscal crisis' of the American government means that without drastic cuts in social expenditures the items at the top of the political agenda cannot be pursued (Domke *et al.* 1983; Pear 1986).

Finally, the New Deal 'compact' between labor unions and the US government has collapsed. This comes at the same time that attempts to incorporate minorities such as blacks and Hispanics have also come to an end. These events increase the likelihood of major social conflict. In a remarkable discussion of contemporary social change Wallerstein (1984: 132) goes so far as to assert that:

Unseated from its role of economic and military hegemony but nevertheless still immensely strong, faced with a relatively declining standard of living, containing an internal class structure in which the correlation of class and ethnicity is extraordinarily high and with strata roughly equal in size, and having developed a pattern of geographical dispersion which concentrates the underclass in the large cities, the US is on a path where it will be difficult in my opinion to avert *de facto* civil war. That is to say, I foresee social unrest that will be far more widespread and violent than what the US knew in the late 1960s.

Conclusion

Fortunately, as the old cliché goes, the future is made, not predicted. 'What Might Happen Otherwise' need not happen. There is no doubt that the Reagan administration has developed policies designed to increase the level of international confrontation and arms production, and to create the basis and support system for intervention around the world (Tucker 1965; Thompson 1985). But the domestic population in 'post-industrial' America must be made to support the very high costs of this growth strategy (Nincic 1985). Despite the box-office success of *Rambo*, the population has been substantially changed by the experience of the 1960s and 1970s. The real question now is whether people can come to see the connection between their domestic impasse and the shifting world-economy. Only by changing how the economy and the polity work at home can confrontation – and war – abroad be averted.

6

Conclusion

This book has pursued two related objectives: to trace the peopling, geographical expansion, economic growth and political development of the United States within the world-economy and to place the present 'impasse' of the United States into a global perspective. As the workings of the world-economy can only be understood in a historical–geographical framework the major unifying theme has been a focus on the evolving regional geography of the United States. In this chapter the general argument is reviewed and weighed against the main alternative of national exceptionalism. Then the argument is presented in a *political* light as a way of demonstrating to Americans that they must learn to live *with* the world if they are to survive *in* it. These are dangerous but challenging times.

From colony to hegemony

Until the 1970s many Americans talked about the special mission of the United States. Such talk has returned again in the 1980s. Undoubtedly this sense of mission, the urge to make the world over in an American image, has been a powerful stimulant of American expansionism from the colonial period to the present. But it has been a variable, whereas expansion has been a constant. It has also been a justification, not simply an element in explanation. Rather than an isolated entity 'cleaving unto itself' until mobilized by the sense of mission, the United States has been a participant from its earliest years in a global system of political and economic relationships in which it has been entwined and to which it has contributed by virtue of its existence as one state in a world of competing states.

The United States was established as an independent state at a time

219

when a European-based global states system was already formed and in the process of development. The date 1648 (the Peace of Westphalia) is often used as a shorthand reference to mark the beginning of the modern European states system. But it was not until the eighteenth century that 'power politics' in the modern sense of the term – associated with the concept of 'foreign policy,' standing and conscript armies drawn from politically mobilized populations, and the 'balance of military power' between member states as an operating principle – emerged as a major feature of global political relations (Hinsley 1966: Chapter 5). This system, however, had been several hundred years in the making and was the product of the expansion of European political and economic activities into the rest of the world.

Under the influence of a tremendous explosion in trade and commerce between the fifteenth and eighteenth centuries, European states emerged that were committed to policies of economic nationalism or 'mercantilism.' For some commentators these are viewed as deriving from the ambitions of political leaders to build the power of their states. To others they represent the hopes of merchants who 'captured' their states in pursuit of their private interests. To the proponents of mercantilism, however, power and wealth were not mutually exclusive goals (Wilson 1967: 495). Each facilitated the other. In the *Traité de l'économie politique*, published in 1615, Montchrestian offered the following set of connections between state-making and trade:

It is impossible to make war without arms, to support men without pay, to pay them without tribute, to collect tribute without trade. Thus the exercise of trade, which makes up a large part of political action, has always been pursued by those people who flourished on glory and power, and these days more diligently than ever by those who seek strength and growth. (Quoted in Tilly 1981: 115)

The dual emphasis on power and wealth should be reiterated. Though states differed in the relative balance of power between independent political leaders and merchants, the survival and prosperity of all of them depended upon the pursuit of the mercantilist formula. The world into which the United States came as an independent state and in which it developed was therefore of a world *political* economy. Even today, and in the face of the globalization of economic activities through new global institutions such as multinational enterprises, the world is still quasi-mercantilist in that every modern state accepts the *responsibility* for managing *its* national economy and regulating *its* international economic relations. Whether states can now carry out these functions *effectively* is another matter entirely.

As a 'member' of this global system, the United States has until recently been consistently competitive and successful. The nature and scale of American expansion have changed dramatically over time. There was nothing predetermined about this. But business and political leaders in the United States proved better able than those of many other states to turn domestic capabilities to international advantage and consequently enhance national wealth and power. How?

In the first place, the United States shifted from a territorial to an 'interactional' mode of expansion in the nineteenth century. The continental expansion of the country was explicitly territorial in nature. By the 1870s the expansion of the United States beyond continental boundaries, however, was bound up with the control of trade and commerce. This shift is symbolized most clearly in foreign policy by the Monroe Doctrine of the early period and the Open Door Policy of the later one (Smith, T. 1981).

The Spanish–American War might seem to contradict this. But even proponents of annexation such as Mahan were clear that the territories resulting from that war were not annexed for the markets and resources they might offer but as strategic points and hinterlands to guarantee access to more lucrative opportunities nearby (Pratt 1936: 13). They were the means of keeping the American foot in the Open Door. In this respect American expansion was taking a leaf out of the British book. For many years the British had tried to minimize their imperial expenditures while maximizing their imperialist returns.

In the second place, the geographical *scope* of American expansion has progressively increased – from continental to hemispheric to global. During the nineteenth century the United States was almost entirely consumed by expansion at continental and hemispheric scales. Only in this century, and more especially since 1945, has the United States become a global actor. As late as 1901 Theodore Roosevelt argued against a global role for the United States. He believed that the United States should dominate the western hemisphere but that elsewhere it should involve itself only in achieving a balance of power (Rosenberg, E. S. 1982: 57). But Roosevelt's successors have had few such reservations. Taft's 'dollar diplomacy' and Wilson's ideal of a world moral order under American stewardship marked the beginning of American expansion into regions of the world previously untouched by American enterprise, influence and culture. Why?

The United States was the first country in a position of significant power within the modern world-economy to view colonial empire as an unnecessary burden rather than as a prerequisite for political–economic dominance. Some writers have traced this development to the aftermath of the Civil War and the emergence of an industrial system that required markets for manufactures and outlets for investment rather than just access to raw materials. The industrial heyday of the United States, beginning in the 1870s and ending, perhaps, in the 1970s, marked a period in which

America had no need of a formal global empire but could increase her imperial power simply by relying upon her superior economy . . . The United States could afford to move from the crude territorial expansionism of Polk to a peaceful and ordered world governed by due respect for property and the Open Door. (Stedman Jones 1983: 233)

The absence of a feudal, aristocratic caste searching for meaningful employment as imperial managers may also have been of some importance (Stedman Jones 1973: 232).

A similar view puts rather more emphasis upon the growth in the United States of an integrated corporate economy and its pervasive commercialization of social and political life (e.g. Trachtenberg 1982; and Rosenberg, E. S. 1982). From this viewpoint colonialism was neither necessary nor desirable. What was important was that American business could have open access to foreign countries without the financial burden of everyday administration and the ideological stigma of colonialism.

Related to this, but missing in most accounts, is the question of why a corporate economy should have developed so successfully in the United States compared to elsewhere. One answer to this brings into focus the link between the nature of the modern state and the growth of capitalism. Meyer (1982: 265) proposes that a relatively 'weak' and decentralized state such as the United States came to have numerous advantages over more bureaucratic and centralized states such as France: (1) centralized states 'tend to mobilize around existing economic possibilities and myths, not emergent ones'; (2) 'the bureaucratic state builds up citizenship and welfare standards, and thus labor costs and productive disadvantages'; (3) 'the bureaucratic state, though capitalist in commitments, may run up too many military costs for its productive base'; and (4) 'the bureaucratic state organizes around a territorial base and economy, rather than strictly around capital accumulation.' After acquiring a territorial base at a continental scale, the American state tended to leave the business of expansion to business except when military activity or intervention became unavoidable. The Founding Fathers' fear of a strong state led to the easy usurpation of public power by private interests (Harrington 1986). This was not without considerable advantage for a time. It delivered hegemony to the United States. However, once pushed too far with the globalization of American business and as the United States took on the roles of the strong state, especially in its competition with the Soviet Union, America's strength turned to weakness.

It is clear, therefore, that American experience has been unique. But this is not in the sense of American ideological exceptionalism. Rather, it is in the context of America's place within a dynamic world-economy. A world-economy (*économie-monde*) has been in existence in at least part of the world since the sixteenth century (Wolf 1983). Today the entire world is involved within the singular division of labor produced by the world-economy. Within this *historical* social system the United States has evolved in its own way. After coming to dominance within the world-economy, indeed, the United States imposed its own mark on the operations of the world-economy. Thus, contrary to national exceptionalists, and also those who would see the United States as a creature of global superorganic imperatives ('the laws of motion of capital', etc. – see Harvey 1982), the changing position of the United States within the world-economy – from colony to hegemony and beyond – has been a product of both 'internal' and 'external' determination. The world-economy has presented certain possibilities, Americans have responded to these and, in turn, forced changes in the world-economy.

Learning to live with the world?

Three different lessons can be drawn from the impasse within which the United States now finds itself within the world-economy. One is to argue that the coalition that brought growth in the past can continue to do so in the future if only the 'invisible hand' of the market can pull the strings no matter what the result. The rhetoric of free trade still appeals powerfully to many politicians and most economists. But there is little reason to believe that market competition and free trade will reverse the relative decline of the US economy. As free traders wait for the pressure of competition to spur US economic growth, governments in other countries are providing investment subsidies, protection from international competition, and support for research and development. If American government policy does nothing more than promote free trade and open markets the US economy is likely to con-

tinue to experience large trade deficits, employment losses in manufacturing industries, and declining relative productivity.

A second lesson is essentially mercantilist. This involves equating
the global 'world-economy' (*économie-monde*) with that of a set of
national economies, the 'world economy' (*économie mondiale*). The
latter, as Wallerstein (1984: 13) has pointed out, presumes that the
world consists of a set of largely independent or 'separate economies'
which are national in scope, and which, under certain circumstances,
'trade with each other.' The consequence of this is to advocate
aggressive and expansionist policies towards the rest of the world
when one's national economy shows signs of decline. Japan, for
example, 'harms the world economy. Since 1980, it has grown slowly
at home and relied on exports for stimulus. It needs to grow faster
domestically so it will import more. But internal growth is retarded
because its outmoded financial system encourages Japanese consumers to save too much' (Samuelson 1985: 86). American economic
vices thus become virtues in a world of competitive national
economies in which American dominance is under threat. At the same
time as blaming the Japanese (and others) for 'unfair' competition, the
liberal economic order that they so successfully exploit must be
defended against its Great Enemy: the Soviet Union. The paradox here
is clear. Mercantilism can perhaps help solve the 'Japanese' (or
economic) problem but it cannot resolve the 'Soviet' (or political–
military) problem. Practicing mercantilism and defending a liberal
economic order are incompatible.

A third lesson rejects the mercantilist solution but also sees the
problem with the let-the-market-decide strategy. This alternative
focuses on a world-economy that has been transformed by the actions
of some Americans (and others) into an entity in which national
divisions have become increasingly irrelevant to business decisions. In
this context the imperative for Americans becomes the possibility of
gaining popular control over their own economy and encouraging and
allowing others to do the same. This does not mean abandoning trade
or global economic relationships, but it does involve giving up what
Williams (1981: 225) calls the 'imperial way of life' with which the
United States has become associated:

The moral of this tale has nothing to do with being soft on communism or
unilateral disarmament or pacifism. We, all of us, here and everywhere, are in
a transition period that offers us the opportunity to imagine and act upon a
way to move on beyond global imperialism to regional communities. Away
from the kind of interdependence programmed by the computers of the multinational corporations to the kind of dialogue that is the substance of a
neighbourhood.

Further reading

There is an immense literature relevant to studying the United States in the context of the world-economy. All that can be done here is to select by chapter the books and essays I have found most useful and which explore key issues in more detail than was possible here.

Chapter 1

Wallerstein (1974; 1979) and Taylor (1985) provide the best overviews of the world-economy perspective. Corbridge (1986) provides a critique of the viewpoint that is very much in line with the approach taken in this book. Evans (1979) and Smith, C. A. (1984) give useful critiques of the world-economy perspective from the point of view of scholars interested in the particular histories of specific places within the world-economy. Brenner (1977) lays out a more general critique from the point of view of a scholar interested in explaining qualitative change in the nature of the world-economy, especially the development of industrialization.

Wolfe (1981a) presents the general argument for America's present 'impasse' within the world-economy. Calleo (1982) and Reich (1983) provide similar arguments but with much less radical prognoses. Harris (1983) proposes a much more 'global' position – to him the 'present crisis' is one of the world-economy pure and simple – but in so doing fails to account for the persisting importance of geopolitical competition between states, a major element in the arguments of Wolfe and Calleo especially (see also Rowthorn 1980). Nevertheless, Harris provides a superb overview of certain critical problems within the world-economy, in particular the trade in armaments and foodstuffs.

Chapter 2

A good discussion of long cycles and their relationship to business cycles of shorter wavelength is provided by Freeman *et al.* (1982). The danger of seeing Kondratieff's long wave as a *cause* rather than as a phenomenon to be explained is discussed in Day (1976). The entire thrust of this chapter is that the *nature* of the world-economy has changed over time and, therefore, that the causes of expansion and contraction have changed also. The history of the world-economy, or of specific places, cannot simply be 'read off' Kondratieff's long cycle; shifts in the cycle must be explained.

Useful overviews of portions of United States history paying close attention to the global context include Williams (1969), Becker and Wells (1984), Calleo and Rowland (1973) and Lafeber (1963). Valuable sources of information on American economic growth and social change are North (1961), Walton and Shepherd (1979), Ward (1971), Weinstein, J. (1968), Fox and Lears (1983) and Davis (1986). On the 'globalization' of the world-economy, see Dunning (1983), Leontief *et al.* (1977), Fröbel *et al.* (1977) and Calleo and Strange (1984).

Chapter 3

There is considerably less literature on the connection between the world-economy and America's regions than on that between the world-economy and the American economy in general. Sources I have found useful include Perloff *et al.* (1960), Billington (1960), Robertson, R. (1973), Archer and Taylor (1981), Dilger (1982), Dorel (1985), Wiley and Gottlieb (1985), Hall (1986) and, above all, Bensel (1984)

Chapter 4

This chapter is based on a large number of sources. The most important ones are Wolfe (1981a), Magaziner and Reich (1982), Bensel (1984), Elazar (1968), Bluestone and Harrison (1982) and Edsall (1984). A number of articles in Sawers and Tabb (1984), *Business Week* and the *New England Economic Review* (published by the Federal Reserve Bank of Boston) were also extremely useful. Burnham (1982) and Sandoz and Crabb (1985) are good sources of information on political and electoral changes in the 1970s and early 1980s.

Chapter 5

Proponents of the post-industrial society scenario usually rely heavily on Bell (1973). Among geographers, Clark, D. (1985) presents the most explicit reference to the post-industrial argument.

The 'opportunity' for economic democracy provided by America's impasse is discussed by Wolfe (1981a). Both Alperovitz and Faux (1982) and Bowles *et al.* (1984) present similar arguments. Keohane (1984) discusses the prospects for a less aggressive and more multilateral involvement of the United States in world affairs 'after hegemony.'

Wallerstein (1984) and Frank (1983) speculate on shifts in global alignments and alliances as a result of increased tension between the United States and Western Europe. Unlike Williams (1981), or the writers noted above, neither sees the present impasse in an optimistic light.

Bibliography

ACIR (Advisory Commission on Intergovernmental Relations) 1980. *Regional growth: a historical perspective.* Washington, DC: Government Printing Office

Ackerman, F. *et al.* 1970. *Income distribution in the United States.* Cambridge, Mass.: URPE

Agnew, J. A. 1981. Homeownership and the capitalist social order. In *Urbanization and urban planning in capitalist society*, ed. M. Dear and A. Scott. New York: Methuen

Agnew, J. A. 1982. Sociologizing the geographical imagination: spatial concepts in the world-system perspective. *Political Geography Quarterly*, 1: 159–66

Agnew, J. A. 1984. Devaluing place: 'people prosperity' versus 'place prosperity' and regional planning. *Society and Space*, 2: 35–45

Alperovitz, G. 1986. No more rich uncle to rich allies. *New York Times*, February 21: A31

Alperovitz, G. and J. Faux 1981. *Rebuilding America: a blueprint for the new economy.* New York: Pantheon

Archer, J. C. and P. J. Taylor 1981. *Section and party: a political geography of American presidential elections, from Andrew Jackson to Ronald Reagan.* Chichester: John Wiley

Archer, J. C. *et al.* 1985. Counties, states, sections, and parties in the 1984 presidential election. *The Professional Geographer*, 37: 279–87

Armington, C. *et al.* 1983. *Formation and growth in high technology businesses: a regional assessment.* Washington, DC: Brookings Institutions

Babson, S. 1973. The multinational corporation and labor. *Review of Radical Political Economics*, 5: 17–25

Bailyn, B. 1955. *The New England merchants in the seventeenth century.* Cambridge, Mass.: Harvard University Press

Balbus, I. D. 1982. *Marxism and domination.* Princeton, NJ: Princeton University Press

228

Barnet, R. J. and R. E. Müller 1974. *Global reach: the power of the multi-nationals.* New York: Simon and Schuster

Batie, S. S. and R. G. Healy 1983. The future of American agriculture. *Scientific American*, 248: 45–53

Becker, W. H. and S. F. Wells (eds.) 1984. *Economics and world power: an assessment of American diplomacy since 1789.* New York: Columbia University Press

Beenstock, M. 1984. *The world economy in transition*, 2nd edn. London: George Allen and Unwin

Bell, D. 1973. *The coming of post-industrial society.* New York: Basic Books

Benjamin, G. L. 1985. The financial stress in agriculture. *Economic Perspectives* (Federal Reserve Bank of Chicago), November/December: 3–16

Bensel, R. F. 1984. *Sectionalism and American political development, 1880–1980.* Madison: University of Wisconsin Press

Bergesen, A. and D. Sahoo 1985. Evidence of the decline of American hegemony in world production. *Review*, 8: 595–611

Bergstrand, J. 1982. The scope, growth, and causes of intra-industry trade. *New England Economic Review*, September/October: 45–61

Berkman, N. G. 1979. Mortgage finance and the housing cycle. *New England Economic Review*, September/October: 54–76

Berry, B. J. L. and F. E. Horton 1970. *Geographic perspectives on urban systems.* Englewood Cliffs, NJ: Prentice-Hall

Billington, R. A. 1945. The origins of middle western isolationism. *Political Science Quarterly*, 60: 44–64

Billington, R. A. 1960. *Westward expansion.* New York: Macmillan

Birch, D. 1979. *The job-generation process.* Cambridge, Mass.: MIT program on neighborhood and regional change

Black Coalition for 1984 1984. *The people's platform.* Washington, DC: Offices of Congressman Walter Fauntroy

Blair, J. M. 1972. *Economic concentration.* New York: Harcourt Brace Jovanovich

Bluestone, B. 1972. Economic crisis and the law of uneven development. *Politics and Society*, 2: 65–82

Bluestone, B. and B. Harrison 1982. *The deindustrialization of America: plant closings, community abandonment, and the dismantling of basic industry.* New York: Basic Books

Boorstin, D. J. 1965. *The Americans: the national experience.* New York: Random House

Bousquet, N. 1980. From hegemony to competition: cycles of the core. In *Processes of the world-system*, ed. T. K. Hopkins and I. Wallerstein. Beverly Hills: Sage

Bowles, S. *et al.* 1984. *Beyond the wasteland: a democratic alternative to economic decline.* New York: Doubleday Anchor

Brenner, R. 1977. The origins of capitalist development: a critique of neo-Smithian Marxism. *New Left Review*, 104: 25–91

Brown, C. J. F. and T. Sheriff 1979. Deindustrialization: a background paper. In *De-industrialization*, ed. F. Blackaby. London: Heinemann

Browne, L. E. 1978. Regional industry mix and the business cycle. *New England Economic Review*, November/December: 35–53

Browne, L. E. 1979. Shifting patterns of interregional migration. *New England Economic Review*, November/December: 17–32

Browne, L. E. 1983. Can high tech save the Great Lake states? *New England Economic Review*, November/December: 19–33

Browne, L. E. 1984. How different are regional wages? A second look. *New England Economic Review*, March/April: 40–7

Browne, L. E. and J. S. Hekman 1981. New England's economy in the 1980s. *New England Economic Review*, January/February: 5–16

Browne, L. E. and R. F. Syron 1979. Cities, suburbs and regions. *New England Economic Review*, January/February: 41–57

Browne, L. E. *et al.* 1980. Regional investment patterns. *New England Economic Review*, July/August: 5–23

Bruchey, S. 1967. *Cotton and the growth of the American economy*. New York: Harcourt Brace Jovanovich

Brummer, A. 1986. American machismo and the sudden darkness. *Guardian Weekly*, February 9: 7

Bunge, W. and R. Bordessa 1975. *The Canadian alternative: survival, expeditions and urban change*. Toronto: Dept. of Geography, York University

Burnham, W. D. 1982. *The current crisis in American politics*. New York: Oxford University Press

Business Week 1980. Revitalizing the American economy, June 30: 56–7

Business Week 1982. Guns versus butter, November 29: 68–85

Business Week 1983. The collapse of world oil prices, March 7: 92–9

Business Week 1983. Why the recovery may skip the farmbelt, March 21: 106–16

Business Week 1983. Time runs out for steel, June 13: 84–94

Business Week 1984. The Northeast: reports of its death have been greatly exaggerated, January 23: 125–8

Business Week 1984. The revival of productivity, February 13: 92–100

Business Week 1984. Uncle Sam could soon be one of the world's biggest debtors, February 27: 106–7

Business Week 1984. The all-American small car is fading, March 12: 88–95

Business Week 1984. Latin America's woes are casting a pall over Miami, April 30: 134–43

Business Week 1984. Illegal immigrants: the US may gain more than it loses, May 14: 126–9

Business Week 1984. Are foreign partners good for US companies?, May 28: 58–9

Business Week 1984. How overseas investors are helping to reindustrialize America, June 4: 103–4

Business Week 1984. The New York colossus, July 23: 98–112

Business Week 1984. Drastic new strategies to keep US multinationals competitive, October 8: 168–72

Business Week 1984. Has the dollar dealt a body blow to American industry?, October 15: 24

Business Week 1986. The hollow corporation: A. The decline of manufacturing threatens the entire US economy, March 3: 57–84; B. The false paradise of a service economy, March 3: 78–81

Calavita, N. 1983. I nuovi equilibri territoriali e demografici negli Stati Uniti. *Storia della città*, 28: 71–82

Calleo, D. P. 1982. *The imperious economy*. Cambridge, Mass.: Harvard University Press

Calleo, D. P. and B. M. Rowland 1973. *America and the world political economy: Atlantic dreams and national realities*. Bloomington: Indiana University Press

Calleo, D. P. and S. Strange 1984. Money and world politics. In *Paths to international political economy*, ed. S. Strange. London: George Allen and Unwin

Carnoy, M. and D. Shearer 1980. *Economic democracy*. New York: Random House

Carr, E. H. 1961. *What is history?* New York: Knopf

Castells, M. 1979. *The economic crisis and American society*. Princeton, NJ: Princeton University Press

Caute, D. 1978. *The great fear: the anti-Communist purge under Truman and Eisenhower*. New York: Simon and Schuster

Chaliand, G. and J.-P. Rageau 1985. *Strategic atlas: comparative geopolitics of the world's powers*. New York: Harper and Row

Chandler, A. D. 1965. *Giant enterprises: Ford, General Motors and the automobile industry*. New York: Harcourt Brace Jovanovich

Checkowoy, B. 1980. Large builders, federal housing programmes, and post-war suburbanization. *International Journal of Urban and Regional Research*, 4: 21–44

Clark, D. 1985. *Post-industrial America: a geographical perspective*, London: Methuen

Clark, V. S. 1929. *History of manufactures in the United States*, Vol. 2. New York: McGraw-Hill

Cloos, G. and P. Cummins 1984. Economic upheaval in the midwest. *Economic Perspectives* (Federal Reserve Bank of Chicago), January/February: 3–14

Cochran, T. C. 1961. Did the Civil War retard industrialization?, *Mississippi Valley Historical Review*, 48: 197–210

Cohen, D. 1972. Does IQ matter?, *Commentary*, April: 17–20

Cohen, J. and J. Rogers 1983. *On democracy: toward a transformation of American Society*. New York: Penguin

Cohen, R. B. 1981. Multinational corporations, international finance, and the

232   *Bibliography*

sunbelt. In *The rise of the sunbelt cities*, ed. D. C. Perry and A. J. Watkins. Beverly Hills: Sage

Collman, C. A. 1931. *Our mysterious panics, 1830–1930*. New York: Morrow

Comisso, E. T. 1979. *Workers' control under plan and market*. New Haven: Yale University Press

Connolly, W. E. 1983. Progress, growth and pessimism in America. *Democracy*, 3: 22–31

Corbridge, S. 1986. *Capitalist world development: a critique of radical development geography*. Totowa, NJ: Rowman and Littlefield

Council of Economic Advisors 1981. *Economic Report of the President*. Washington, DC: Government Printing Office

Cox, K. R. 1973. *Conflict, power and politics in the city: a geographic view*. New York: McGraw-Hill

Cox, R. W. 1981. Social forces, states and world order. *Millenium*, 10: 126–55

Davis, M. 1980. Why the US working class is different. *New Left Review*, 123: 3–46

Davis, M. 1981. The New Right's road to power. *New Left Review*, 128: 28–49

Davis, M. 1986. *Prisoners of the American dream: politics and economy in the history of the US working class*. London: Verso

Day, R. B. 1976. The theory of the long cycle: Kondratiev, Trotsky, Mandel. *New Left Review*, 99: 67–82

DeGrasse, R. W. 1984. The military: shortchanging the economy. *Bulletin of the Atomic Scientists*, May: 39–43

Denison, E. F. 1979. *Accounting for slower economic growth: the United States in the 1970s*. Washington, DC: Brookings Institution

Derber, M. 1970. *The American idea of industrial democracy, 1865–1956*. Urbana: University of Illinois Press

Destler, C. M. 1946. Entrepreneurial leadership among the 'Robber Barons': a trial balance. *Journal of Economic History*, 6 (Supplement): 28–49

de Tocqueville, A. 1966. *Democracy in America*, trans. George Lawrence. New York: Harper and Row

Dickerson, O. M. 1951. *The Navigation Acts and the American Revolution*. Philadelphia: University of Pennsylvania Press

Dilger, R. J. 1982. *The sunbelt/snowbelt controversy: the war over federal funds*. New York: New York University Press

Domke, W. K. *et al.* 1983. The illusion of choice: defense and welfare in advanced industrial democracies, 1948–1978. *American Political Science Review*, 77: 19–35

Dorel, G. 1985. *Agriculture et grandes entreprises aux Etats-Unis*. Paris: Economica

Dorn, W. L. 1940. *Competition for empire, 1740–1763*. New York: Harper

Drinnon, R. 1980. *Facing West: the metaphysics of Indian-hating and empire-building*. Minneapolis: University of Minnesota Press

Dunning, J. H. 1983. Changes in the level and structure of international production: the last one hundred years. In *The Growth of International Business*, ed. M. Casson. London: George Allen and Unwin

Easterlin, R. A. 1961. Regional income trends, 1840–1950. In *American Economic History*, ed. S. Harris. New York: McGraw-Hill

Ecker, D. S. and R. F. Syron 1979. Personal taxes and interstate competition for high technology industries. *New England Economic Review*, September/October: 25–32

Edsall, T. B. 1984. *The new politics of inequality*. New York: Norton

Edwards, C. 1985. *The fragmented world: competing perspectives on trade, money and crisis*. London: Methuen

Elazar, D. J. 1968. Megalopolis and the new sectionalism. *Public Interest*, 11: 67–85

Elazar, D. J. 1972. *American federalism: a view from the states*. New York: Crowell

Ellis, R. P. 1979. How have regional energy patterns changed since 1973?, *New England Economic Review*, March/April: 62–79

Erdevig, E. 1984. The bucks stop elsewhere: the Midwest's share of federal R and D. *Economic Perspectives* (Federal Reserve Bank of Chicago), November/December: 13–23

Evans, P. 1979. Beyond center and periphery: a comment on the contribution of the world system approach to the study of development. *Sociological Inquiry*, 49: 15–20

Fainstein, S. S. and N. I. Fainstein 1976. The federally-inspired fiscal crisis. *Society*, 13: 27–32

Farnsworth, C. 1983. Third World debts mean fewer jobs for Peoria. *New York Times*, December 11: E3

Farnsworth, C. 1984. The too-mighty dollar takes a toll of manufacturing jobs. *New York Times*, September 23: E3

Feagin, J. R. 1984. Sunbelt metropolis and development capital: Houston in the era of state capitalism. In *Sunbelt/snowbelt*, ed. L. Sawers and W. K. Tabb. New York: Oxford University Press

Feagin, J. R. 1985. The social costs of Houston's growth. *International Journal of Urban and Regional Research*, 9: 164–85

Fieleke, N. S. 1981a. Challenge and response in the automobile industry. *New England Economic Review*, July/August: 37–48

Fieleke, N. S. 1981b. Productivity and labor mobility in Japan, the United Kingdom and the United States. *New England Economic Review*, November/December: 27–36

Fitzgerald, F. 1972. *Fire in the lake: the Vietnamese and the Americans in Vietnam*. Boston: Little, Brown

Flynn, P. 1984. Lowell: a high technology success story. *New England Economic Review*, September/October: 39–49

Focus 1984. Poverty in the United States. Where do we stand now?, *Focus* (Institute for research on poverty, University of Wisconsin-Madison), 7, 1

Fogel, R. W. and W. L. Engerman 1974. *Time on the cross: the economics of American Negro slavery*, Vol. 1. Boston: Little, Brown

Fox, E. W. 1971. *History in geographic perspective: the other France*. New York: Norton

Fox, R. W. and T. J. J. Lears (eds.) 1983. *The culture of consumption: critical essays in American history, 1880–1980*. New York: Pantheon

Frank, A. G. 1983. *The European challenge*. Nottingham: Spokesman Books

Franko. K. L. 1980. *European industrial policy: past, present and future*. Brussels: The Conference Board in Europe

Freeman, C. *et al.* 1982. *Unemployment and technical innovation: a study of long waves and economic development*. London: Frances Pinter

Frickey, E. 1942. *Production in the United States, 1860–1914*. Cambridge, Mass.: Harvard University Press

Fröbel, F. *et al.* 1977. *Die neue internationale Arbeitsteilung*. Reinbeck: Rowolt

Fusfield, D. R. 1968. Fascist democracy in the United States. In *Conference papers of the Union for Radical Political Economics*. New York: URPE

Galbraith, J. K. 1955. *The great crash*. Boston: Houghton Mifflin

Gallman, R. E. and E. S. Howle 1971. Trends in the structure of the US economy since 1840. In *The reinterpretation of American economic history*, ed. R. W. Fogel and S. L. Engerman. New York: Harper and Row

Genetski, R. J. and Y. D. Chin 1978. *The impact of state and local taxes on economic growth*. Chicago: Harris Bank

Genovese, E. 1972. *Roll, Jordan, roll: the world the slaves made*. New York: Pantheon

Gilbert, F. 1961. *The beginnings of American foreign policy*. Princeton, NJ: Princeton University Press

Glad, P. W. 1964. *McKinley, Bryan, and the people*. Philadelphia: Lippincott

Goodman, R. 1984. *The last entrepreneurs: America's regional wars for jobs and dollars*. Boston: South End Press

Goodrich, C. 1960. *American promotion of canals and railroads, 1800–1890*. New York: Harper and Row

Goodwin, W. 1965. The management center in the United States. *Geographical Review*, 55: 1–16

Goodwyn, L. 1978. *The Populist moment: a short history of the Agrarian Revolt in America*. New York: Oxford University Press

Gottdeiner, M. 1985. *The social production of urban space*. Austin: University of Texas Press

Goulden, J. 1976. *The best years, 1945–1950*. New York: Atheneum

Guerrari, P. and P. C. Padoan 1986. Neomercantilism and international economic stability. *International Organization*, 40: 29–42

Gutman, H. G. 1976. *The black family in slavery and freedom, 1750–1925*. New York: Pantheon

Haavind, R. 1983. The US is losing the technology race. *High Technology*, June: 115–19

Hall, A. R. 1968. *The export of capital from Britain: 1870–1914*. London: Methuen

Hall, T. D. 1986. Incorporation in the world-system: toward a critique. *American Sociological Review*, 51: 390–402

Harper, L. A. 1942. Mercantilism and the American Revolution. *American Historical Review*, 23: 1–15

Harper's 1984. The myth of high-tech jobs. August: 24

Harrington, M. 1976. *The twilight of capitalism*. New York: Simon and Schuster

Harrington, M. 1986. *The dream of deliverance in American politics*. New York: Knopf

Harris, N. 1983. *Of bread and guns: the world economy in crisis*. London: Penguin

Hartz, L. (ed.) 1964. *The founding of new societies*. New York: Harcourt Brace

Harvey, D. 1982. *The limits to capital*. Chicago: University of Chicago Press

Hawley, E. 1966. *The New Deal and the problem of monopoly*. Princeton, NJ: Princeton University Press

Heady, E. O. 1962. *Agricultural policy under economic development*. Ames, Iowa: Iowa State University Press

Hendrickson, P. 1985. The fields of fear. *Guardian Weekly*, June 16: 17

Hershey, R. D. 1985. Savings rate in US lowest since 50s despite incentives. *New York Times*, October 29: A1, D23

Hershey, R. D. 1986. Spending rose sharply in 'Reagan Revolution.' *New York Times*, February 2: A20

Hill, R. C. 1984. Economic crisis and political response in the Motor City. In *Sunbelt/snowbelt*, ed. L. Sawers and W. K. Tabb. New York: Oxford University Press

Hinsley, F. H. 1966. *Sovereignty*. New York: Basic Books

Hobsbawm, E. 1969. *Industry and empire*. London: Penguin

Hofstadter, R. 1971. *America at 1750: a social portrait*. New York: Knopf

Hogan, M. J. 1985. American Marshall planners and the search for a European neocapitalism. *American Historical Review*, 90: 44–72

Hollingsworth, J. R. 1963. *The whirligig of politics*. Chicago: University of Chicago Press

Howe, I. (ed.) 1984. *Alternatives: proposals for America from the Democratic left*. New York: Pantheon

Huntington, S. P. 1982. American ideals versus American institutions. *Political Science Quarterly*, 97: 1–37

International Association of Machinists and Aerospace Workers 1984. *Let's rebuild America*. Washington, DC: International Association of Machinists and Aerospace Workers

International Monetary Fund 1979. *The direction of trade.* Washington, DC: IMF

Jacoby, L. R. 1972. *Perception of air, noise and water pollution in Detroit.* Ann Arbor: University of Michigan Dept. of Geography

Jenks, L. H. 1927. *The migration of British capital to 1875.* New York: Knopf

Jensen, M. (ed.) 1965. *Regionalism in America.* Madison: University of Wisconsin Press

Jones, R. C. (ed.) 1984. *Patterns of undocumented migration: Mexico and the United States.* Totowa, NJ: Rowman and Allenheld

Josephson, M. 1962. *The robber barons: the great American capitalists, 1861–1901.* New York: Harcourt Brace Jovanovich

Jusenius, C. and L. C. Ledebur 1977. *Where have all the firms gone? An analysis of the New England economy.* Washington, DC: Government Printing Office

Keohane, R. O. 1984. *After hegemony: cooperation and discord in the world political economy.* Princeton, NJ: Princeton University Press

Key, V. O. 1964. *Politics, parties and pressure groups,* 4th edn. New York: Crowell

Kiernan, V. G. 1974. *Marxism and imperialism.* New York: St Martin's Press

Kindleberger, C. P. 1978. *Manias, panics, and crashes: a history of financial crises.* New York: Harper and Row

Kirby, J. T. 1984. The South as a pernicious abstraction. *Perspectives on the American South,* 2: 167–80

Kolakowski, L. 1978. *Main currents in Marxism,* Vol. 3. New York: Oxford University Press

Kolko, G. 1962. *Wealth and power in America.* New York: Harper and Row

Kondratieff, N. 1984. *The long-wave cycle.* New York: Richardson and Snyder

Kuttner, B. 1983. The free trade fallacy. *New Republic,* March 15

Kuznets, S. 1945. *National product in wartime.* New York: National Bureau of Economic Research, Publication No. 44

Lafeber, W. 1963. *The new empire: an interpretation of American expansion, 1860–1898.* Ithaca: Cornell University Press

Lakshmanan, T. R., W. Anderson and M. Jourabchi 1984. Regional dimensions of factor and fuel substitution in US manufacturing. *Regional Science and Urban Economics,* 14: 381–98

Landes, D. S. 1969. *The unbound prometheus.* Cambridge: Cambridge University Press

Lappé, F. M. 1983. Sweden's third way to worker ownership. *The Nation,* February 19: 203–4

Lappé, F. M. and J. Collins 1978. *Food first: beyond the myth of scarcity.* New York: Random House

Leontief, W. 1985. After manufacturing, what?, *Population Today,* 13: 2

Leontief, W. *et al.* 1977. *The future of the world economy.* New York: Oxford University Press

Lernoux, P. 1984. The Miami connection: Mafia/CIA/Cubans/banks/drugs. *The Nation*, February 18: 186–98

Leven, C. 1981. Regional shifts and metropolitan reversal in the US. Paper presented at the Conference on Urbanization and Development, June 1–4, Vienna

Lewis, A. 1985. Angry at success: looking at Tokyo instead of ourselves. *New York Times*, August 12: A17

Libby, O. 1984. The geographical distribution of the vote of the thirteen states on the Federal Constitution, 1787–1788. *Bulletin of the University of Wisconsin*, 1: 7–116

Lindblom, C. 1978. *Politics and markets*. New York: Basic Books

Lindsey, R. 1985. Layoffs by high technology concerns raising fears in California Valley. *New York Times*, November 10: A5

Lipset, S. M. 1966. *The first new nation*. New York: Doubleday Anchor

Little, J. S. 1985. Foreign direct investment in New England. *New England Economic Review*, March/April: 48–57

Lively, R. A. 1955. The American system: a review article. *Business History Review*, 29: 81–96

Lubell, S. 1952. *The future of American politics*. New York: Harper

Luger, M. I. 1984. Federal tax incentives as industrial and urban policy. In *Sunbelt/snowbelt*, ed. L. Sawers and W. K. Tabb. New York: Oxford University Press

Luria, D. and J. Russell 1984. Motor City changeover. In *Sunbelt/snowbelt*, ed. L. Sawers and W. K. Tabb. New York: Oxford University Press

Lynd, S. 1970. Beyond Beard. In *Towards a new past: dissenting essays in American history*, ed. B. J. Bernstein. London: Chatto and Windus

Lynd, R. and H. Lynd 1937. *Middletown in transition*. New York: Harcourt Brace and World

MacLaughlin, J. and J. A. Agnew 1986. Hegemony and the regional question: the political geography of regional industrial policy in N. Ireland 1945–1972. *Annals, Association of American Geographers*, 76: 247–61

Magaziner, I. and R. B. Reich 1982. *Minding America's business: the decline and rise of America's economy*. New York: Vintage Press

Main, J. T. 1965. *The social structure of revolutionary America*. Princeton, NJ: Princeton University Press

Makler, H. *et al.* 1982. *The new international economy*. Beverly Hills: Sage

Malecki, E. J. 1980. Dimensions of R and D location in the United States. *Research Policy*, 9: 2–22

Mandel, E. 1975. *Late capitalism*. Atlantic Highlands, NJ: Humanities Press

Martineau, H. 1837. *Society in America*, Vol. 1. London: Saunders and Otley

Mayer, M. 1978. *The builders*. New York: Norton

McDermott, J. 1982. The secret history of the deficit. *The Nation*, August 21–8: 129, 144–6

McIntyre, R. 1986. Funding mergers and higher salaries. *New York Times*, February 23: Business Section, 2

McKinney, J. A. and K. A. Rowley 1985. Trends in US high technology trade. *Columbia Journal of World Business*, 20: 69–81

McManus, M. 1976. How the Northeast finances southern prosperity. *Empire State Report*, 2, 9: 347–52

Melman, S. 1974. *The permanent war-economy: American capitalism in decline.* New York: Simon and Schuster

Merk, F. 1963. *Manifest destiny and mission in American history: a reinterpretation.* New York: Knopf

Meyer, J. W. 1982. Political structure and the world-economy. *Contemporary Sociology*, 11: 263–6

Miller, W. 1949. American historians and the business elite. *Journal of Economic History*, 9: 184–208

Mills, C. W. 1945. The American business elite: a collective portrait. *Journal of Economic History*, 5: 20–44

Mintz, A. and A. Hicks 1984. Military Keynesianism in the United States, 1949–76. *American Journal of Sociology*, 90: 411–17

Mintz, S. W. 1985. *Sweetness and power: the place of sugar in modern history.* New York: Viking

Modigliani, F. 1985. How economic policy has gone awry . . . *New York Times*, November 3: Business section, 2

Montgomery, D. 1978. On Goodwyn's populists. *Marxist Perspectives*, Spring: 167–72

Moore, Jr. B. 1966. *Social origins of dictatorship and democracy.* Boston: Beacon Press

Mosley, H. G. 1985. *The arms race: economic and social consequences.* Lexington, Mass.: Lexington Books

Moynihan, D. P. and F. Mosteller (eds.) 1972. *Equality of educational opportunity.* New York: Random House

Musgrave, P. 1975. *Direct investment abroad and the multinationals: effects on the United States economy.* Washington, DC: Government Printing Office

Newman, J. 1984. *Growth in the American South: changing regional employment and wage patterns in the 1960s and 1970s.* New York: New York University Press

New York Times 1979. Drilling more but finding less. National Economic Survey, January 7: 43

New York Times 1985. New studies show growth rate for clerical jobs is starting to slow down, October 7: A23

Nincic, M. 1985. The American public and the Soviet Union: the domestic context of discontent. *Journal of Peace Research*, 22: 345–57

Noble, K. B. 1985. Big strikes found on decline in US. *New York Times*, July 12: A13

Noble, K. B. 1986. Study finds 60% of 11 million who lost jobs got new ones. *New York Times*, February 7, A1, A15

North, D. C. 1961. *The economic growth of the United States, 1790 to 1860.* Englewood Cliffs, NJ: Prentice-Hall

Northeast–Midwest Institute 1981. *National and state energy expenditures, 1970–80.* Washington, DC: Northeast–Midwest Institute

Nozick, R. 1974. *Anarchy, state and Utopia.* New York: Basic Books

O'Connor, J. 1973. *The fiscal crisis of the state.* New York: St Martin's Press

OECD (Organization for Economic Cooperation and Development) 1961–81. *National accounts, 1961–1981.* Paris: OECD

Olmsted, F. L. 1861. *Journey in the back country.* New York: Porter

Olson, M. 1982. *The rise and decline of nations: economic growth, stagflation, and social rigidities.* New Haven: Yale University Press

Owsley, F. 1949. *Plain folk of the old South.* Baton Rouge: Louisiana State University Press

Parboni, R. 1981. *The dollar and its rivals: recession, inflation and international finance.* London: Verso

Parboni, R. 1985. *Il conflitto economico mondiale.* Milan: Itas

Parry, J. H. 1971. *Trade and dominion: the overseas European empires in the eighteenth century.* New York: Praeger

Pear, R. 1986. Chipping away at the idea of entitlement. *New York Times,* February 9: E8

Peet, R. 1983. Relations of production and the relocation of United States manufacturing industry since 1960. *Economic Geography,* 59: 112–43

Perlman, S. 1928. *A theory of the labor movement.* New York: Macmillan

Perloff, H. S. *et al.* 1960. *Regions, resources, and economic growth.* Lincoln, NE: University of Nebraska Press

Piven, F. F. and R. Cloward 1972. *Regulating the poor.* New York: Vintage

Pollack, A. 1986. Service jobs start to drift abroad, too. *New York Times,* March 23: Section 12, 7

Potter, D. M. 1965. *People of plenty.* Chicago: University of Chicago Press

Potter, J. 1965. The growth of population in America, 1700–1860. In *Population in history,* ed. D. V. Glass and D. E. C. Eversley. London: Edward Arnold

Pratt, J. W. 1935. The ideology of American expansion. In *Essays in Honor of William E. Dodd,* ed. A. Craven. Chicago: University of Chicago Press

Pratt, J. W. 1936. *Expansionists of 1898: the acquisition of Hawaii and the Spanish Islands.* Baltimore: Johns Hopkins University Press

Pratt, J. W. 1957. *Expansionists of 1812.* Gloucester, MA: Peter Smith (reprint of 1925 edn, Macmillan, NY)

Pred, A. R. 1974. *Major job-providing organizations and systems of cities.* Washington, DC: Association of American Geographers Resource Publications, No. 27

Priestly, H. L. 1939. *France overseas through the old regime: a study of European expansion.* Englewood Cliffs, NJ: Prentice-Hall

Rawls, J. 1971. *A theory of justice.* Cambridge, Mass.: Harvard University Press

Reed, C. 1985. The rage of Rambo. *Guardian Weekly*, August 4: 20

Rees, J. and B. L. Weinstein 1983. Government policy and industrial location. In *United States public policy: a geographical view*, ed. J. House. Oxford: Clarendon Press

Reich, R. B. 1983. *The next American frontier: a provocative program for economic renewal.* New York: Penguin

Reinhold, R. 1985. Texas boom in people, as in oil, slows. *New York Times*, October 22: A6

Republican Party 1980. *National platform.* Washington, DC: Republican Party National Committee

Riche, R. W. *et al.* 1983. High technology today and tomorrow: a small slice of the employment pie. *Monthly Labor Review*, November: 50–8

Robbins, W. 1985a. Shakeout in farming: the dilemma of the banks and the core of the losses. *New York Times*, July 19: A19

Robbins, W. 1985b. Farms' crisis endangering rural towns. *New York Times*, October 14: A1, A16

Robertson, J. O. 1980. *American myth, American reality.* New York: Hill and Wang

Robertson, R. 1973. *History of the American economy*, 3rd edn. New York: Harcourt Brace Jovanovich

Rogers, E. M. 1985. *The high technology of Silicon Valley.* College Park, MD: University of Maryland Institute for Urban Studies

Rogin, M. P. 1975. *Fathers and children: Andrew Jackson and the subjugation of the American Indian.* New York: Knopf

Rohatyn, F. G. 1984. *The twenty-year century.* New York: Random House

Rones, P. L. 1980. Moving to the sun: regional job growth, 1968 to 1978. *Monthly Labor Review*, March: 25–32

Rose, H. M. and D. R. Deskins 1980. Felony murder: the case of Detroit. *Urban Geography*, 1: 1–21

Rosenbaum, D. E. 1986. Investment found unaided by tax shift. *New York Times*, February 11: D8

Rosenberg, E. S. 1982. *Spreading the American dream: American economic and cultural expansion, 1890–1945.* New York: Hill and Wang

Rosenberg, N. 1967. Anglo-American wage differences in the 1820s. *Journal of Economic History*, 27: 221–9

Rossiter, C. 1953. *Seedtime of the Republic.* New York: Random House

Rowthorn, B. 1980. *Capitalism, conflict, and inflation.* London: Lawrence and Wishart

Russett, B. 1970. *What price vigilance?* New Haven: Yale University Press

Sadler, G. E. 1982. *Comparative productivity dynamics: Japan and the United States.* Houston: American Productivity Center

Sale, K. 1975. *Power shift: the rise of the Southern rim and its challenge to the Eastern establishment.* New York: Vintage

Salsbury, S. 1962. The effect of the Civil War on American industrial development. In *The economic impact of the American Civil War,* ed. R. Andreano. Cambridge, Mass.: Schenkman

Sampson, A. 1977. *The arms bazaar: from Lebanon to Lockheed.* New York: Viking

Sampson, A. 1981. *The money lenders.* London: Hodder and Stoughton

Samuelson, R. J. 1985. Our Japan obsession. *Newsweek,* August 12: 56

Sandoz, E. and C. V. Crabb (eds.) 1985. *Election '84: landslide without a mandate?* New York: Mentor

Sanger, D. E. 1985. Prospects appear grim for US chip makers. *New York Times,* October 29: D1

Sawers, L. and W. K. Tabb (eds.) 1984. *Sunbelt/snowbelt: urban development and regional restructuring.* New York: Oxford University Press

Saxenian, A. L. 1984. The urban contradictions of Silicon Valley: regional growth and the restructuring of the semiconductor industry. In *Sunbelt/snowbelt,* ed. L. Sawers and W. K. Tabb. New York: Oxford University Press

Saxenian, A. L. 1985. Review 3: 'Let them eat chips.' *Society and Space,* 3: 121–8.

Sayer, A. 1984. *Method in social science: a realist approach.* London: Hutchinson

Schlesinger, A. M. 1945. *The age of Jackson.* Boston: Little, Brown

Schlosstein, S. 1984. *Trade wars: greed, power, and industrial policy on opposite sides of the Pacific.* New York: Congdon and Weed

Schmidt, W. E. 1986. Not all of the South is in the sunbelt. *New York Times,* February 9: E8

Schneider, K. 1985. Financing US deficit abroad. *New York Times,* November 7, D1, D8

Schumpeter, J. A. 1946. The decade of the twenties. *American Economic Review,* 36: 1–10

Semiconductor Industry Association 1983. *The effect of government targeting on world semiconductor competition: a case history of Japanese industrial strategy and its costs for America.* Cupertino, Calif.: Semiconductor Industry Association

Shannon, F. A. 1945. *The farmer's last frontier, 1860–1897.* New York: Farrar and Rinehart

Shepherd, J. F. and G. M. Walton 1972. Trade, distribution, and economic growth in colonial America. *Journal of Economic History,* 32: 128–45

Shepherd, J. F. and G. M. Walton 1976. Economic change after the American revolution: pre-war and post-war comparisons of maritime shipping and trade. *Explorations in Economic History,* 13: 397–422

Smith, C. A. 1984. Local history in global context: social and economic transitions in Western Guatemala. *Comparative Studies in Society and History*, 26: 193–228

Smith, D. 1979. *The public balance sheet*. Washington, DC: Conference on Alternative State and Local Policies

Smith, D. E. 1979. *Where the grass is greener: living in an unequal world.* Baltimore: Johns Hopkins University Press

Smith, H. R. 1955. *Economic history of the United States*. New York: Ronald Press

Smith, T. 1981. *The pattern of imperialism: the US, Great Britain and the late industrializing world*. Cambridge: Cambridge University Press

Soule, G. 1968. *Prosperity decade: from war to depression, 1917–1929*. New York: Harper and Row

Stedman Jones, G. 1973. The history of US imperialism. In *Ideology in social science: readings in critical social theory*, ed. R. Blackburn. New York: Pantheon

Syron, R. F. 1978. Regional experience during business cycles: are we becoming more or less alike?, *New England Economic Review*, November/December: 25–34

Tabb, W. K. 1984. Urban development and regional restructuring, an overview. In *Sunbelt/snowbelt*, ed. L. Sawers and W. K. Tabb. New York: Oxford University Press

Tagliabue, J. 1985. Italian lira stabilized as trading continues. *New York Times*, July 23: D1

Tarbell, I. 1936. *Nationalizing of business, 1878–1898*. New York: Macmillan

Taylor, P. J. 1985. *Political geography: world-economy, nation-state and locality*. London: Longman

Thernstrom, S. 1964. *Poverty and progress*. Cambridge, Mass.: Harvard University Press

Thomas, H. and C. Logan 1982. *Mondragon: an economic analysis*. London: George Allen and Unwin

Thomas, R. P. 1968. British imperial policy and the economic interpretation of the American revolution. *Journal of Economic History*, 27: 436–40

Thompson, K. W. 1985. Commentary. In *Intervention and the Reagan doctrine*, ed. K. W. Thompson. New York: The Council on Religion and International Affairs

Thurow, L. C. 1980. *The zero-sum society*. New York: Basic Books

Thurow, L. C. 1984. America's banks in crisis. *New York Times Magazine*, September 23: 48, 72–7, 108–9

Thurow, L. C. 1986. The hidden sting of the trade deficit. *New York Times*, January 19: F3

Tilly, C. 1981. *As sociology meets history*. New York: Academic Press

Trachte, K. and R. Ross 1985. The crisis of Detroit and the emergence of global capitalism. *International Journal of Urban and Regional Research*, 9: 186–217

Trachtenberg, A. 1982. *The incorporation of America: culture and society in the gilded age*. New York: Hill and Wang

Tucker, R. W. 1985. Intervention and the Reagan doctrine. In *Intervention and the Reagan doctrine*, ed. K. W. Thompson. New York: The Council on Religion and International Affairs

Turner, F. J. 1896. The problem of the West. *Atlantic Monthy*, 78: 289–97

Turner, F. J. 1920. *The frontier in American history*. New York: Holt

Turner, F. J. 1932. *The significance of sections in American history*. New York: Holt

Tyson, L. D. 1980. *The Yugoslav economic system and its performance in the 1970s*. Berkeley: University of California, Institute of International Studies

United Nations 1978. *UN demographic yearbook*. New York: United Nations

US Bureau of the Census 1957. *Historical statistics of the United States*. Washington, DC: Government Printing Office

US Bureau of the Census 1975. *Statistical abstract of the United States*. Washington, DC: Government Printing Office

US Bureau of the Census 1981. *Statistical abstract of the United States, 1981*. Washington, DC: Government Printing Office

US Department of Commerce 1985. *Survey of current business*, July

US Department of Energy 1980. *Residential energy consumption and expenditures, April 1978 through March 1979*. Washington, DC: DOE/EIA

US Department of Transportation 1981. *The US automobile industry*. Washington, DC: Government Printing Office

Vaughan, R. 1977. *The urban impacts of federal policies*. Santa Monica, Calif.: Rand Corporation

Vaughan, R. 1979. *State taxation and economic development*. Washington, DC: Council of State Planning Agencies

Vaupel, J. W. and J. P. Curhan 1969. *The making of multinational enterprise*. Cambridge, Mass.: Harvard University Graduate School of Business

Vaupel, J. W. and J. P. Curhan 1974. *The world's multinational enterprises*. Geneva: Centre for Education in International Management

Vernon, R. 1966. International investment and international trade in the product cycle. *Quarterly Journal of Economics*, 80: 190–207

Vidal, G. 1986. Requiem for the American empire. *The Nation*, January 11: 1, 15–19

Vining, D. R. 1982. Migration between the core and the periphery. *Scientific American*, 247: 44–53

Virtue, G. O. 1936. Capitalistic aspects of the colonial economy. In *Explorations in economics*, ed. G. O. Virtue. New York: Harper

Walker, R. A. 1978. The transformation of urban structure in the nineteenth century and the beginnings of suburbanization. In *Urbanization and conflict in market societies*, ed. K. Cox. Chicago: Maaroufa Press

Walker, R. A. 1985. Is there a service economy? The changing capitalist division of labor. *Science and Society*, 49: 42–83

Wallerstein, I. 1974. *The modern world-system I: capitalist agriculture and the origins of the European world-economy in the sixteenth century*. New York: Academic Press

Wallerstein, I. 1979. *The capitalist world-economy*. Cambridge: Cambridge University Press

Wallerstein, I. 1984. *The politics of the world-economy: the states, the movements, and the civilizations*. Cambridge: Cambridge University Press

Wallerstein, I. *et al.* 1979. Cyclical rhythms and secular trends of the capitalist world-economy: some premises, hypotheses and questions. *Review*, 2, 4: 483, 500

Walton, G. M. 1971. The new economic history and the burdens of the Navigation Acts. *Economic History Review*, 2nd ser., 24: 533–42

Walton, G. M. and J. F. Shepherd 1979. *The economic rise of early America*. Cambridge: Cambridge University Press

Walzer, M. 1983. *Spheres of justice*. New York: Basic Books

Ward, D. 1971. *Cities and immigrants: a geography of change in nineteenth-century America*. New York: Oxford University Press

Wasylenko, M. and T. McGuire 1985. Jobs and taxes: the effect of business climate on states' employment growth rates. *National Tax Journal*, 38: 497–512

Wayne, S. J. 1980. *The road to the White House: the politics of presidential elections*. New York: St Martin's Press

Weinstein, B. L. and R. E. Firestine 1978. *Regional growth and decline in the United States*. New York: Praeger

Weinstein, B. L. and H. Gross 1985. The frost belt's revenge. *Wall Street Journal*, November 19: 1

Weinstein, B. L. *et al.* 1985. *Regional growth and decline in the United States*, 2nd edn. New York: Praeger

Weinstein, J. 1968. *The corporate ideal in the liberal state 1900–1918*. Boston: Beacon Press

White, M. 1985. The dough turns sour in America's breadbasket. *Guardian Weekly*, February 24: 9

Wilber, C. K. and K. P. Jameson 1983. *An inquiry into the poverty of economics*. Notre Dame, Ind.: Notre Dame University Press

Wiley, P. and R. Gottlieb 1985. *Empires in the sun: the rise of the new American West*. Tucson: University of Arizona Press

Wilkins, M. 1970. *The emergence of multinational enterprise: American business abroad from the colonial era to 1914*. Cambridge, Mass.: Harvard University Press

Williams, W. A. 1969. *The roots of the modern American empire*. New York: Vintage

Williams, W. A. 1981. *Empire as a way of life: an essay on the causes and character of America's present predicament along with a few thoughts about an alternative.* New York: Oxford University Press

Wilson, C. H. 1967. Trade, society and the state. In *Cambridge economic history of Europe*, Vol. 4. Cambridge: Cambridge University Press

Wilson, J. Q. 1975. A guide to Reagan country: the political culture of Southern California. In *The ecology of American political culture: readings*, ed. D. J. Elazar and J. Zikmund II. New York, Crowell

Wolf, E. R. 1983. *Europe and the people without history.* Berkeley: University of California Press

Wolfe, A. 1981a. *America's impasse: the rise and fall of the politics of growth.* Boston: South End Press

Wolfe, A. 1981b. Sociology, liberalism, and the radical right. *New Left Review*, 128: 3–22

Wood, G. W. 1966. Rhetoric and reality in the American Revolution. *The William and Mary Quarterly*, 23: 1–32

Woodworth, W. *et al.* (eds.) 1985. *Industrial democracy: strategies for community revitalization.* Beverly Hills, Calif.: Sage

Worster, D. 1986. *Rivers of empire: water, aridity and the growth of the American West.* New York: Pantheon

Young, M. E. 1958. Indian removal and land allotment: the civilized tribes and Jacksonian justice. *American Historical Review*, 64: 31–45

Zelinsky, W. 1964. *The cultural geography of the United States.* Englewood Cliffe, NJ: Prentice-Hall

Zunz, O. 1982. *The changing face of inequality: urbanization, industrial development and immigrants in Detroit, 1880–1920.* Chicago: University of Chicago Press

Index